The New Boundaries
between Bodies and Technologies

The New Boundaries
between Bodies and Technologies

Edited by

Bianca Maria Pirani and Ivan Varga

Cambridge Scholars Publishing

The New Boundaries between Bodies and Technologies, Edited by Bianca Maria Pirani and Ivan Varga

This book first published 2008 by

Cambridge Scholars Publishing

15 Angerton Gardens, Newcastle, NE5 2JA, UK

British Library Cataloguing in Publication Data
A catalogue record for this book is available from the British Library

ISBN (10): 1-84718-499-5, ISBN (13): 9781847184993

TABLE OF CONTENTS

PREFACE

ROBERTO CIPRIANI

Body and Society

There is a peculiarity of the human body that makes it an entity in between something natural and which perpetuates itself following the basic rules of common life: human body can reproduce itself on and on (as it has been happening since some million years, as witnessed by the female hominid fossil, named Lucy, which was found in East Africa, and early hominids which date back to 6-8 million years), and for some more years it will keep on reproducing, thus giving way to the most meaningful among human activities.

Given such premises, the subject of the following essays concerns the metaphoric value of some expedients which give importance to the body in its relation to new technologies, taking into consideration the fluid boundaries between bodies and ultra-modern technologies.

A medium such as the camera can easily preserve dead bodies in life with the re-presentation or iconic representation of their human profiles, no more as social actors but as indirect protagonists, so to become a sort of social bound and referring examples for (in the name of) their ideas and main issues. In a sense, the passed away do not really 'pass away' thanks to the newborn vitality that comes out from printing the images of their life.

As a matter of fact, as Francesco Faeta reminds us (2000: 163), 'the pictures of the "passed away" state, in their objectivity, a mutation and a closing in time experience, because they are an operating activity and hold parts of the essence and constitution of the subject, therefore they are a good means for eidetic reconstruction'.

The 20th Century was defined as the century of the body according to the major attention paid to all the aspects of corporeality, so to make of it a sort of religion of the body. Besides, the media and the new means of diffusion for the works of art have highly contributed to the increase of such a new phenomenon.

Also on a linguistic level there is something undefined and vague in
the Italian language: the term '*salma*' (corpse) is a highly undetermined
one because it can be '*soma*' (again an imprecise word which refers to the
old Greek term for "body") in the sense of the carried luggage but also to
the corpse (object of cultural rites), to the dead body, without neglecting
the possibility of remanding to the living body as an imperfect part
because of its materiality. On another hand, speaking of '*salmerie*'
(baggage train) means referring to useful materials for survival both from
a 'grocery' point of view (eating supplies) and from a military point of
view (for munitions, both for defence and salvation).

An historical and sociological approach to the body
In France, more than in Italy, the issue of the body was widely
explored, as witnesses the major work of Alain Corbin, Jean-Jacques
Courtine and Georges Vigarello (2006) on *L'Histoire du corps* (The
History of the Body). Some symbolic and tragic reflections in the
meantime are those of Primo Levi (1976) in *Se questo è un uomo* (If This
Is a Man), where he says, in 1947: 'My body is not my body anymore'.
And Varlam Salamov (1995) in *I racconti della Kolyma* (Stories from
Kolyma) seems to continue the sentence by writing: 'the bones are
freezing, the brains and the soul start stiffing'. Both witness the experience
of a dictatorship, respectively the German Nazism and Soviet
Communism. Both ideologies consider the body as centre of their sense of
reality, completely based on rhetoric, supremacy and triumph of power,
which have martyrized with unbelievable cruelty helpless people guilty of
disapproving dominant ideas or belonging to an 'undesirable' race.

The contemporary trends are different. However, today as well there is
an overestimation of the health and powerful body always followed by an
humiliation of the body itself, (and even if it does not reach the violence
and trials of German concentration camps and that of forced residence and
hard labour camps in Russia imposed by GULAG) giving hard proofs to
bodies piercing them through, or with perforations and any other kind of
harming, thus denying them and considering them abstract instead of
material, then passing through them as if without resistance or consistency.
This kind of treatment has known no limits and has mummified before its
time the body, labelling it as a mail package, branding it as an animal. It is
the body exposed and humiliated.

Apart from such examples of disrespect for human body and its
vulnerable human nature, the true decision maker of the body is the DNA.
With its helicoidally trend tending to left (nobody has still explained why),

it is *de facto* capable of pre-determining diseases and eventual fatal exitus according to a pre-defined and not elidible schedule.

The denied body

The body is often named, invoked, designated, desired, but it is as well blamed, annihilated, denied. The paradox is that the body is at the same time scrutinized, stenographed, enlarged, examined also in the inner part where it was never before possible to enter, *in interiore hominis*, which means in its inner parts. From radiography to endoscope, these are technologies which examine the body in all its elements inscrutable to the naked human eye. This way, scientific discoveries which render all this possible are highlighted and celebrated, as a profound analysis that was hardly believable even a few years ago.

In the meantime, however, the human body is at times banned, or prayed, according to the different ideological trends of secular rationality and earthly immanence, but also on the basis of the religious and confessional perspectives, that can be diverse and opposed to the secular perspectives.

Anthropology has made the body a major subject. Painting, sculpture as well as photography and cinematography have loosened it from tinsels, clothes, costumes that were hiding it. It was made extremely visible, becoming a new object of attention and of ethic-religious discussions. The homosexual movement has led the discussion to the extreme consequences of exhibition that meant as a revenge and provocation with political implications especially in the *Homosexual Front for Revolutionary Action*.

In another aspect, we have to notice as well that nowadays the body is more and more repairable, modifiable, and adaptable. Prosthesis substitutes what lacks or what does not work. '*Soma*' (body) is therefore re-programmed. Even death can be postponed, at least partially, thanks to the intervention of intubations, breathing-help technologies, oxygen suppliers. Such 'prolonged-deaths' have important examples in the last days of the Spanish dictator Francisco Franco, of the Brazilian Politician Tancredi Neves, and of Suharto in Indonesia.

Moreover, from a moral point of view, there are issues ranging from eugenics to euthanasia. This means from life to death, endlessly along the life course. But also less dramatic issues are concerned, such as cosmetics, dietetics, plastic surgery, and ultraviolet radiations for untimely tanning, tattoos, piercing, or any kind of mutilations.

From an iconic point of view, the exhibited body has passed from the old barber's calendars (now pieces for exhibitions in museums) given for free to clients, to the present calendars for men, women or homosexuals, also sold for charity.

We have to remember that the opposite of beautiful, graceful and well-done is highly successful in the arts. Some part of literature, cinema, artistic works, television shows deal with or show what is around of monstrous, hideous, anomalous, corrupt, ugly, beastly and fierce having a great success among people of all ages. This way, the difference and the multiform as the possible impossibility become focus of interest, the booster for commercial actions, and medium for new trends and fashions with traces of ideology and intellectual tendencies.

We have quickly passed from the numbers tattooed on the arms of prisoners in the concentration camps, to the personal use of the skin as a manifesto for exhibiting one's feelings or ideas.

The visibility of the body turns into athletic performance, exhibition, show, that has its apex in the dynamic of torture as a proof that challenges the capacity of resistance of the actor and the subject, but also of the watchers, who can be just occasional watchers or the main instigators. Fakirs act in such a way that strongly provoke the actors implying them at all levels; holding on to the insensitivity acquired in troubled and painful situations, fakirs can cause sceptical attitudes or, on the contrary, feelings of deep respect as well as a sort of veneration, as known by Hindu practising the extreme forms of pain annulment.

New dynamics of the body

A recent text by Hervé Juvin (2006) seems to celebrate the utmost glorification of the human body. New aesthetic plastic surgery, new genetics, new dietetic regimes and new aesthetics have given an important tribute to the ideology of wellness.

In the meantime alternative solutions are being studied, so to give new energy and development to the body industry: issues such as abortion and contraceptives are problems of the past. The new issues are cloning, homosexual marriages, gene banks, artificial insemination, embryo modifications.

Medical assistance is extremely diffused. The body is considered an economic property *sui generis,* sold in parts for transplants, lent for bone marrow transfer (at times, for the same reason,   procreation for medical purposes, e.g. stem-cell research, can be performed). In other words, the body is slowly becoming one among the large number of commercialized products. The body is, however, our own product, a private resource, that is giving way to a new form of capitalism: the use of private property in relation to our body (with evaluations in terms of prizes for each single organ, like kind of a new butcher's window for human pieces, where prizes and items are catalogued as in an insurance contract for compensation for damages).

The present tendency seems to preserve people that are not capable of reproducing itself for age reasons. The search of wellness at all costs keeps relentlessly going on.

The conscience of having a body

In the history of human life, many different attitudes towards the body were alternatively shown; a special attention has been paid to its more vital expression: the blood. In this regard, the writing of Piero Camporesi (1984) on the 'juice of life' is a good example. The blood is only a component of a more complex whole, it is the object of strong and long-lasting feelings expressed at best in the work of arts, which leave deep impacts between generations: the famous *David* by Michelangelo to the one by Donatello, who inspired it; from *Adam* at the Sistine Chapel to the beautiful Riace's bronzes, where the movement of the bodies is translated into elegance, allusions, references, symbols, that propose   wide possibilities for reflection on the meaning of life.

Only later, with Rosmini (1926) was the doctrine of the corporal feeling affirmed, according to which any knowledge starts from the body perception. According to Rosmini, founder of a religious congregation, the starting point is the consciousness of one's own corporeity that allows feeling life in all its expressions, therefore, to become aware of the external events. Thanks to the body we can perceive what is external and foreign. The feeling of belonging derives from the body as well as the conception of oneself and of the other. Therefore, experience as *Erlebnis* is linked to the consciousness of the body as *Leib,* and totally different from the nature of the body as *Körper.* The first (consciousness of the body) differs from the second (the nature of the body) in its relation to the degree of consciousness and reflexivity.

Again, corporeality offers an extraordinary supply of meaning. Corporeality is language, meaning, message, in one word it is culture and one can experience it through gesture, communication, intelligence, awareness and consciousness.

Therefore, this question is not useless: 'Do we have a body or rather are we a body?' As a matter of fact, the body is the only real property of human beings for a quite long time. The body is the centre and the meaning of all and it gives sense to the world around.

Finally, from a sociological point of view the most important issue is the social aspect of the body, that is expressed both by sexuality and fasting or diets, as well as by physical exercises, and as a carrier of social interactions and feelings.

References

Camporesi P. (1984), *Il sugo della vita. Simbolismo e magia del sangue* (The Juice of Life. Symbolism and Magic of Blood), Milano: Comunità.

Corbin A., Courtine J.- J., Vigarello G. (2006), *Histoire du corps* (The History of the Body), 3 vols, Paris. Seuil.

Faeta F. (2000), *Il santo e l'aquilone. Per un'antropologia dell'immaginario popolare nel secolo XX* (The Saint and the Kite. For an Anthropology of the Popular Imagination in the 20th Century), Palermo: Sellerio.

Juvin H. (2006), *L'avènement du corps* (The Coming of the Body], Paris: Gallimard.

Levi P. (1976), *Se questo è un uomo*, Einaudi, Torino (English translation, "If This Is a Man", Harmondsworth: Penguin Books, 1979).

Rosmini S. A. (1926), *Nuovo saggio sull'ordine delle idee* (A New Essay on the Order of Ideas), Milano: Sodalitas Salamov V. (1995), *I racconti della Kolyma* (Stories from Kolyma), Milano: Adelphi.

INTRODUCTION

BIANCA MARIA PIRANI & IVAN VARGA

The invention and proliferation of microelectronic technologies and the rapid pace of their constant development and application – mostly in the developed world – introduced a new phase not only in the role of technologies in human's life but also brought about serious consequences for almost all aspects of the individual's life and social relations.

We refer to those technologies that are now integrated into everyday life. It also deeply affects the human body. The sensory and cognitive processes undergo a profound change as the authors of this book analyze them. As well, the concept and flow of time also acquire a new meaning. It would not be an exaggeration to compare these developments with the discovery of quantum mechanics. Heisenberg (1959:156) stated the discovery of quantum mechanics changed the concept of reality. He also said that after quantum mechanics reality is not any longer the one experienced by perceptions. Heisenberg was of the view that in quantum theory the law *tertium non datur* (no third possibility) had to be replaced by that of *complementarity*.

The physical world and electronic virtual world are not separate, as much current discussions might lead one to believe; in fact they are intricately intertwined. The present analyses address the links between social constructions of the human body and the growth of completely mmersive realities (known as Virtual Reality or VR) constructed trough computer software. Human bodies form a basis for social relationships. Although a VE (virtual environment) minimizes ambulatory experience, users interacting with virtual technologies constitute material phenomena engaged in practices. Experiments with users wearing Head Mounted Displays (HMDs) confirm that such technologies exercise important effects on human bodies and brains.

We need to consider the roles of virtual places as well as physical ones, of electronic connections as well as asynchronous encounters and transactions in addition to synchronous ones. Therefore we have now a new economy of presence within which we continually choose among the possibilities of synchronous and asynchronous communication, presence and telepresence.

The mobility of humans not only tremendously increased but also took another dimension. Since the 1990s mobility has emerged as an important cross-disciplinary research agenda referred to as the *mobility turn* and the new mobilities paradigm. The slippery and intangible nature of mobility makes it an elusive object of study. Yet mobility is central to what it is to be human. It is a fundamental spatial feature of existence and, as such, provides a rich terrain from which narratives can be and have been constructed.

From the first kicks of a newborn baby to the travels of international business people, mobility is everywhere. It plays a central role in discussions of the body and society (Cresswell, 1999, 24, 175, 192). Culture is no longer tided to places, but is hybrid, dynamic: more about routes than roots. The social is no longer seen as bound by customary rules but as caught up in a complex array of twenty-first century mobilities (Urry, 2000). Philosophy and social theory look to the end of sedentarism and the rise of new forms of nomadism (Deleuze and Guattari, 1986).

Mobility, then, is more central to both the world and our understanding of it than ever before. And yet mobility itself, and what it means, remains unspecified. It is a kind of blank space that stands as an alternative to place, boundedness, foundations and stability.

This 'blank space' is the interface between chaos and order that embodies the ways in which messages move from the material to the cultural. In this process, these messages are themselves transformed: they shift from being energies, matters, objects, into thoughts, ideas, cultural artefacts. According to Serres (1995c: 293), 'angels in their multitude are a better metaphor for getting a grip on the circulations and connections of multifarious, heterogeneous entities: humans, knowledge, languages, objects, processes'. Sherry Turkle, a professor in MIT program on Science, Technology, and Society and a leading researcher in this area, observes in her book, *The Second Self* (1984) that computers, and particularly the Internet, are evocative technologies, bringing out latent habits of mind and body. According to her, people relate to computers in two major patterns: master and cooperation. Mastery is seeing the computer as a microworld that one can control in all its aspects. Cooperation is conceiving of the computer as an ally in creating, communicating, and doing other tasks. In *Life on the Screen* (1995), Turkle argues that the Internet gives rise to a postmodern personality that is supplanting the modern persona that has dominated technological culture for the last few centuries. While the modernist personality is unitary, the postmodern persona is more flexible and playful: it is a 'body on the move'. Contradictory aspects can coexist comfortably in a single *bricolage*, because such complexity reflects a postmodern understanding of reality and our limited ability to map it.

'Bodies on the move', therefore, are both rule and exception. They are the rule because the body movements do not go beyond the physical confines. They are the exception because microelectronic technologies allow a mobility that is not determined any longer by physical factors. Namely, it is the movement of the mind (operating in the virtual space) that does not require any longer physical displacement.

While classical sociology and earlier modern sociologists dealt with structural changes in society, they did not analyze the human body as such. Today, however, sociology has to pay attention to the impacts that microelectronic technologies and the spread of cyberspace cause on the human body. It is always difficult to decide whether the body produces order or disorder, norm-bound regularity or anomalies. These two definitions: 'the body as a temporary site' and an 'ordering machine' are of course not unrelated; in order to be identified as a location, the body must distinguish between an inner and an outer, the self and the other, and this takes place through installing a specific principle of order. This brings us to the problem of identifying the body as linked to a particular life form. Namely, new technologies can and do modify the body. No longer do sociologists complain about the lack of attention paid to the role of the body in social processes. There are multiple intellectual endeavours examining the role of the body in a number of ways. For instance, Brian Turner (1994, 1996) identifies three areas where the body has been treated in some details: the body's symbolic significance as a metaphor for social relationship; as a necessary component in the analysis of gender, sex, and sexuality; and in the context of the study of medical issues. However, as Turner notes, within these various enterprises, the body and embodiment remain illusive and ill-defined. Turner provides the broad outlines of what a general sociological theory of the body would require: a complex account of the idea of embodiment which can incorporate the 'systematic ambiguity of the body as corporeality, sensibility and objectivity' (1994: X-XI); a conceptualization of the social actor as embodied attached to an analysis of how the 'body image functions in social space' (1994: 11). It perceives the body as a thoroughly historical and cultural entity. This outline, highly plausible as it is, omits the discussion on the relation between bodies, embodiment and technologies.

John O'Neill, the noted Canadian sociologist, in his book *Five Bodies: Re-figuring Relationships* (O'Neill, 2004) explores the relationship between the human body and the social institutions, and introducing the concept of *communicative body* states that it is 'the general medium of our world, of its history, culture and political economy.' (2004:4) He also emphasizes that the natural (biophysical) body is intrinsically coupled with the symbolic meanings every society attaches to it. Bourdieu's concept of *habitus*, too, includes considerations about the body and interpersonal, social communication as well as its symbolic presentation.

A key task of this book is to analyze the complexity of the connections between nature, society and technologies by problematizing the corporeal body as vehicle for the interrelational dynamics amongst creative, informational, physical and virtual processes and flows that are currently reshaping the so called 'mobility turn'. It would seem that there is a process of disembodiment: certain functions are lifted out of a particular body and invested in a particular technology. Furthermore, however, we might suggest that, within Western tradition, there is a more general ethos of disembodiment. The idea of disembodiment is attached to a notion of progress. This ethos of disembodiment seems always to be undermined by what we call the 'return of the body'. The point is that it always seems to come to depend upon the body in some guise or other. Many functions of body parts are most obviously delegated to technology. For example, the zip is a convenience because it saves on the complex and repetitive manipulation of fingers and thumbs that was once necessary to use hook and eye fastenings. Machinery and lately robots substitute for human muscle power and recently, computers, together with artificial intelligence (AI) programmes substitute for brain power. The biological, social and machine – or biosoma, for short – paradigm emphasizes the complex nature of a city, which involves these three components, and their interaction with the environment. The machine component encompasses artifacts, from infrastructure to houses from industry to vehicles, from computers to clothing.

Bringing to the extreme the human—machine interactions, David Levy, a renowned expert in artificial intelligence, in his latest book *Love + Sex with Robots: The Evolution of Human--Robot Relationships* (Levy, 2007) predicts that by 2050 both men and women would 'enjoy physical and emotional bonds with life-like, apparently conscious and remarkably suave robots.' (Cf. his interview with Siri Agrell, *The Globe and Mail*, 15. November 2007.) Levy, tracing the developments in constructing artificial replicas of natural objects – beginning as early as in the first century B.C.E. when Heron of Alexandria 'constructed some water-powered mechanical birds, entire flocks of them, that even emitted realistic chirping sounds created by a water-driven device.' (2007:3) Later, in the 18th

Century, the fascination with artifacts imitating natural objects was widespread. In the 20th Century, with the development of AI and industrial robotics there were two important results: the repetitive tasks done by robots have the same quality and are not influenced by human errors; as well, the use of robots can substitute tasks in the household normally done by human physical efforts (e.g. vacuuming, lawn mowing).

However, robots endowed with artificial intelligence represent a totally new form, a new age inasmuch as Levy and like-minded scientists claim, the human-machine interaction would include affective, emotional elements.[1]

While it is not entirely implausible that human beings could evelop an affective relationship to objects (as to a car), it is highly questionable whether – as it happens in real life – being attracted to, or falling in love with, a particular person for his or her appearance, intellect, wit, taste, (even smell, as recent research indicates), etc. could take place. In other words, we find it highly questionable whether robots could develop the great variety of human feelings and sentiments as experienced in embodied beings. At very least, in order to do justice to the complexity of the con-nections between nature, society and technology, what is needed is a collective effort at breaching the disciplines (very generally, the natural and social sciences). Incorporating human bodies into the understanding of social relations can make us better comprehend the ways in which the human is not simply 'tied to' the social, but also 'tied to' the complex web of the technological and the natural. Such an approach provides a broader, albeit continually shifting material base from which theory can develop.

As the introduction of quantum theory in physics revolutionized the concepts of the physical world, the emergence of the cyberworld (or cyborg) revolutionized the communicative processes, mobility and the concept and practices of the body. One could even say that the micro-electronic technologies and the accompanying discoveries (e.g. electronic imagining technologies) have a more direct effect on people's everyday life.

This book, therefore, focuses on the emerging field of 'mobility research' by problematizing the corporeal body as vehicle for inter-relational dynamics amongst imaginative, informational, physical and virtual forms of mobility.

---

[1] John von Neumann, one of the founding fathers of cybernetics, in his posthume book, *Theory of Self-Reproducing Automata* (1966) has already in the early 1950s elaborated the mathematical model of computers that would be capable to improve themselves, thus imitating evolution. However, his model did not involve emotions, feelings, sentiments.

Our bodies are where we locate individual differences: my body is this, yours is another one. They, however, also share operational exchanges: for example, I can sympathize with you and help when you are ill, because I expect the same in a similar situation. Nevertheless, the establishment and operation of spatial differences amongst bodies is unpredictable and relational to other bodies and places. While special transformations are important, in the broader context of social transformation, space rather expresses than reflects society. For instance, the degree of constraints on spatial mobility expresses the openness or lack of it of a society. Also the types of habitat in urban conglomerates are closely related to social stratification. Therefore, space is a fundamental dimension of a society, inseparable from the processes of social organization and social change.

Both the Internet and mobile telephony (which now includes text messages and transmission of pictures) make possible new forms of communicating on the move, at least in developed societies.[2]

One can communicate with places and people around the globe without even leaving one's own home. Thus, physical mobility is complemented or can even be replaced by electronic communication, including tele-conferencing, which enables face-to-face communication over space and time. As well, forms of social life become critical sites in which the very organization of space and society are in a state of flux. From a sociological perspective, Barry Wellman (1999:23) argues that the informational city is built around this double system of communication. Accounting for mobilities in the fullest sense challenges social science to change both the objects of its inquiries and the methodology of research. A particular problem requiring additional studies is how the proliferation of micro-electronic technologies affects everyday life. The same people who extensively use this technology are still living their lives; therefore, they have emotions and relations with their loved ones and assume different roles in society and family. In other words, the question is whether this technology could dominate or even conquer the individual's life.

Proximity and connectivity are imagined in new ways – often enhanced by communication devices and very likely being 'on the move' – and this transforms the home and affects the human body. More and more homes are connected by electronic devices to the outside world, privacy is endangered by obtaining personal data through fraudulent or government-sanctioned means and, though they are just preliminary results, MRI scannings indicate that the brain shows distinct reactions during high-tech

---

[2] Although mobile telephony is overtaking the conventional telephone network in developing societies, mainly because it is cheaper to install than build landlines.

communication processes. Also, observing people's extensive use of mobile devices (phones, iPods, BlackBerries) in public spaces one can see the
impact of the proliferation of electronic communication. The spread and
variety of explicit adult web-sites also indicate the expansion of vir-tual
sex. Owing to the public nature of microelectronic communication, it is
not impossible that the concept of intimacy is undergoing changes. As we
argued earlier, the emotive life of human beings could also be affected: it
is questionable whether emotions conveyed by electronic media are e-
quivalent to the ones expressed by embodied persons.

Apart from the aforementioned phenomena, there is a change in sensory perception. Among the five senses only two, sight and hearing, are
used in microelectronic communication. It remains to be seen whether this
would cause deterioration of the remaining ones. On the other hand, technologies derived from, or enabled by, microelectronic technologies (dia-
gnostic imagination or computer assisted medical procedures, tele-medical
procedures, such as helping physicians or nurses in remote communities
performing complicated tasks, finding medical advice on the Net etc.) are
beneficial for the health of both body and mind.

This spreading and multi-dimensional space is virtual, densely webbed
and complex. A vast realm is being accessed through mediations of our i-
maginative and technical representations and at the same time the existence of space flows is changing our perception of space, time and
interactivity. Places move closer in time; time becomes instantaneous
('real time') and social links become decentered. While this entails many
advantages, such as immediate access to a vast amount of knowledge, data
and information, instantaneous communication, etc., it also creates an illu-
sionary reality. Namely, what users of ITCs perceive as reality is not what
surrounds them in their everyday life. One does not experience physical
pain or have his/her tooth extracted by ITCs. One can see scenes – war,
celebrations, crowdedness, and so forth – without actually being exposed
to them. Admittedly, it is possible to empathize with their real actors, but
this differs from the lived experience.

We have to consider the cognitive consequences of the technological
advances and increased mobility. The iconic culture penetrated the culture
and learning of late-modern or post-modern societies (suffice it to think of
the increasing amount of pictures in social science textbooks, including
university ones, at least in North America). It is highly questionable whether this development contributes to the ability of abstract, conceptual
thinking, or rather the opposite is true. It is not our intention to minimize
the advantages the new technologies have brought about in acquisition of
factual knowledge. On the contrary, one can say that they have contributed

to the democratization of knowledge since anyone with a computer can find data, facts and opinions on just about everything. However, these facts do not foster problem-solving abilities or abstract thinking. We are not making value judgments but, rather, we are pointing out probable directions that could counteract the afore-mentioned process of demo-cratization of knowledge. Namely, the solutions presented make people rely upon anonymous authorities whose opinions are often accepted without critical thoughts. Of course, this phenomenon existed prior to the widespread use of computer based knowledge, but the growing reliance on this type of knowledge promotes an uncritical acceptance of the data and opinions provided.

There is another element in the cognitive field: the discrepancy between the scientific and technical insights into the nature of things and the everyday life experience of people. One does not have to know the physics and mechanisms of internal combustion engine to drive a car, nor has to know the codes of a computer program in order to use it (moreover, an average user utilizes only a fraction of the full capacity of the machine). These examples could serve as a prolegomenon to a much more complex problem, namely, the discrepancy between the actual life experience and the ever more intricate scientific explanations of the phenomena experienced. Ilya Prigogine, in his short book, *Temps à devenir. A propos de l'histoire du temps* (1994), states that people seek *certitude* in science. And, we can add, in technology as well. The image of the physical universe, however, does not lend itself to certitude.

Prigogine (1994:10-11) uses the example of *time* and explains that according to many modern theorists of physics 'time as natural successive unfolding does not exist. The universe is there but does not 'become'. The apparent arrow of time is therefore an illusion that ought to be surpassed, eliminated'. According to him, 'our experience of existence is based on time, on the difference between past and future. This is our par excellence existential dimension. We become, we are not' (here Prigogine uses the Hegelian distinction between *Werden* and *Sein*. However, our everyday life experience includes the present as well. Similarly, the directional nature of the time arrow is felt, and all societies mark the passing of time by periodic holidays or festivities, either religious or secular (such as national holidays, carnivals or carnival-like feasts). Prigogine, however, emphasizes that in the 20th century the scientific discoveries have changed the concept, and says:

> I believe that now one has to believe in a *time arrow*, in an evolving universe, in a universe where there are not only laws but a universe in which there are also events, as in history. There are laws and events. And

the reasons for this radical change of attitudes are unexpected. The first one is the discovery, about thirty years ago, of *the constructive role of time* (ibid. 27-28).

He points to the discovery of rhythmic phenomena in the molecular world and adds that a living being is a whole of rhythms as the rhythms of heart, hormones, brain waves, cell divisions, etc. This is possible because the living being is not in a state of equilibrium. The non-equilibrium is the most extraordinary way nature has invented for coordinating the phenomena and make possible complex phenomena (ibid. 29).[3]

To research into the effects on human beings using digital technologies, the challenge is to learn how the 'natural' perception of time (the time arrow) is disrupted by the instant communication ('real time' bridging or even eliminating the organism's responses to the flow of time) and how the rhythms of the body and mind developed during the evolutionary process are effected by modern technologies. There are initial evidences that there is, indeed, an impact on the body and mind rhythms but much more research is needed. What is obvious is that communication of human beings is undergoing a substantive change, and the disembodiment in the communication process is advanced.

It is also observable that the innate creative potential of the mind is being rapidly replaced by technological devices. We have already mentioned the proliferation of means of instant communication. We also see more and more youngsters using iPods, spending more time in front of a computer. And now, toy companies are bringing to the market toys that are pre-programmed to imitate human responses (e.g. talking Barbie dolls) without requiring children to deploy their own imagination. One does not necessarily have to accept all ideas of Huizinga about *homo ludens* (the playful man) without abandoning concerns about the ability to spontaneity of the incoming generations.

Let us return to the concept of 'mobilities'. It encompasses not only large-scale movements of people, goods, services and capital across the globe, but also the more local processes of daily commuting, moving through public space as well as the movement of material objects within everyday life. Whether it is too little or too much or of the wrong sort at the wrong time, the issue of movement is central to many lives, organizations or governments. Ideas of 'hyper-mobility' and 'instant communication' drive contemporary business strategy, advertising and govern-

---

[3] Actually, Norbert Wiener and the early theoreticians of cybernetics came to similar conclusions, in particular concerning the directions (forward and reverse) thermodynamic takes in nature.

ment policies as well as academic exchanges but also elicit strong political critiques by those who are marginalized or harmed by these developments. Many public, private or non-governmental organizations seek to understand, monitor, manage and transform aspects of these multiple mobilities as well as the new 'immobilities': the social exclusion and eventual threats to security of groups.

Many academics also express concerns about the changing relationship between the body, society and technologies. For instance, Andrew Webster in his article 'Innovative Health Technologies and the Social: Redefining Health, Medicine and the Body' (2002) airs certain qualms about the intrusive nature of latest medical technologies and says that 'new health technologies may promise more than is deliverable and make new demands on those who try to deliver them.' (2002:444.)

We ought to re-formulate and re-interpret 'mobility'. Traditionally it meant physical displacement of human bodies. It certainly did not disappear. On the contrary, both immigration, i.e. looking for jobs in geographic areas different from the traditional home place, and tourism, i.e. temporary movements for entertainment, enjoyment or broadening one's horizons, are expanding. However, there is also an 'instant mobility' in cyberspace, which entails a tension, or even a contradiction between virtual mobility and physical immobility (for instance, a currency trader can carry out transactions around the globe when sitting at his/her desk).

There were instances in social theory that treated 'the move' – downward mobility and marginalization – as an anomaly (in the 1950s mobility studies, especially in the USA, usually dealt with upward mobility) happening in the fringes of society. However, in contemporary society this 'exception' does not apply anymore.

Therefore, the first part of the book entitled *Mobility Nodes and Spatial Mobilities* deals with the fluidities of what Zygmunt Bauman (2000) has called 'liquid modernity'.

Donna Haraway (1989: 49) wrote that in a world marked by rapid, startling innovations in information technology, electronic communications and biological engineering, 'the boundary between science fiction and social reality is an optical illusion'. This statement has been occasionally criticized for being an exaggeration, but it is right to say that some projections in science fictions, in particular the extrapolations concerning the transformative capabilities of technology and their effects on humans, have become reality. In this sense there is not a great difference between science fiction and utopia, which Ernst Bloch defined as the 'not-yet-here'. Ilya Prigogine and Isabelle Stengers (1984: 310) said 'Being and Becoming are not to be opposed one to another: they express two related aspects of the reality'. Our lives are increasingly transformed in ways, and by devices, that seem to emerge from the pages of speculative fiction.

There is, however, a great difference between science fiction and technological reality. The latter is the result of purposive rational thinking that underlies scientific thought, respectively is built upon previous technological achievements. At the same time, the main tendency of these changes seems to transform the world into a virtual reality-game. How do identities, bodies and real spaces become remade by the textual, aural and visual technologies of screens and keypads that populate mobile phones, Internet cafés and PC-based virtual communities and cities? In Virtual Environments (VEs) users interacting with technologies that create VEs form the material element, engaged in practices. Users wearing Head Mounted Displays (HMDs) confirm a sense that technologies, such as virtual reality, are capable to have a grip on human bodies. A virtual online common, like absolute space, would offer an infinitely extensible grid for the potential re/unification of separate individuals, with plenty of room for commerce as well. Such a digital 'public sphere' could permit i-maginative vaulting of Norbert Elias's modern wall surrounding the individual. Nevertheless, the critical spatial separation of user's bodies in 'absolute space' would remain unaltered. Each spatially isolated, distinct individual would become a discrete modern category onto himself/herself. VEs – vision disguised as space – are the ideal public sphere for imaginative subjectivities who believe that 'they' have been virtually freed of bodily constraints. Virtual Environments constitute a privileged psychic variation on contemporary human's homelessness. Divorced from the body's constraining intelligence, the 'fully extensible self' busies itself with the fantasy of building a virtual home in a 'post-symbolic' environment.

At the same time this self embraces the digital means that serve to extend its own psychic disarticulation in the hope to sharpen its ability to control personal meaning by living the virtual experience both as leisure and culture. However, this same person undergoes the biological process of ageing, could suffer from disease, experience the joy over the birth of a child, savour a good meal, etc., etc. – in other words he or she is living in a real, non-virtual world as well.

The challenge for social sciences and psychology is to examine the dialectical relationship in the contemporary world between natural life processes, the reality of everyday life and the virtual reality that is creeping into, and often dominating, people's lives. Similarly, there is a contradiction between the value-neutral nature of the digital technologies and their use. It can foster ordinary people's participation in the political process by organizing support for a cause or a political party or a candidate for president. On the other hand, it could send messages of a criminal nature (e.g. child pornography) or help organizing terrorist activities.

We live therefore in a fragmented society in which distinctions between nature and culture, biology and politics, law and transgression, mobility and immobility, reality and representation are increasingly blurred.

Thus, far from the 'borderless world' (Ohmae, 1990) and the 'end of geography', owing to the economic advantages arising from ICTs, the current uneven access suggests that existing inequalities will be intensified amongst and within countries, socially and spatially, and while stages of development may be leapfrogged, uneven development will remain.

Despite these stark contrasts, there has been a rapid increase in Internet access in some of the poorer countries, especially in Brazil, China and Malaysia, as cyber centres or corridors are being developed. These centres bridge the digital divide between richer and poorer countries, but simultaneously create even starker spatial and social divisions and contrasts internally as people and places are selectively connected and bypassed. The development of a new multimedia supercorridor in Malaysia indicates how rapidly poorer countries can move directly to the technological frontier.

The digital divide between nations coexists with complex divisions within countries by location, social class, gender and ethnicity. A new urban dualism is emerging from the opposition between 'the spaces of flows that links places at a distance on the basis of their market value, their social selection, and their infrastructural superiority' and 'the space of places' that isolates people in their neighborhoods as a result of their low incomes and lack of connections' (Castells 2001: 241).

New technologies may assist social change but it is by no means guaranteed. Thus while the technologies have progressive potential and in principle could transcend the social relations of its creation, their full progressive capacity remains constrained by the existing uneven social relations. Therefore, at present, while enormous changes are taking place in the lives and bodies of people in different places that are becoming increasingly entwined on a global scale as a consequence of the new technologies, social and spatial divisions are still widening.

Contemporary information and communication technologies have transformed many people's lives in positive ways and have the potential for transforming many more. New ICTs and other technologies possess great potential for productivity increases and transforming bodies and lives. ICTs allow knowledge and information to spread around the world much more quickly than ever before and understanding of its effects is a necessary condition for change. Yet, it is not a sufficient condition. From an academic perspective new technologies raise questions about appropriate disciplinary boundaries, especially the separation of studies between the developed and less developed world when economic and social processes and people, bodies and places are increasingly interconnected. Thus, the framework proposed in the second part of this book, *The New Boundaries between Bodies and Technologies*, provide some indications of how this might be done. It attempts to grasp, through specific examples, the changes that result in changing the boundaries between temporalities of the bodies and microelectronic technologies. The chapters provide a comparative analysis of the conception of technologies as the work of the 'mind' over matter that contradicts the modern orientation which dominated modern culture up until the second part of the 20th century. It also reshapes the modern bodies by the electronic technologies. The 'technologization' of the bodies that we are experiencing today is a new phase in the existence of technology.

This book shows how scientific knowledge both embeds and is embedded in social identities, institutions, representations and discourses. Accordingly, the authors argue, ways of knowing the world are inseparably linked to the ways in which people seek to organize and control it. Trough studies of emerging knowledge, research practices and political institutions, the authors offer fresh analytic perspectives on the nexus of knowledge, power and technologies. They show that cyberspace is no mere virtual reality but a rich geography of practices and power relations. Above all, they offer an overview of how top scientists currently understand the processes connecting conscious experience with the world around us.

The authors and editors are fully aware of the immense amount of research to be done with the aim to demonstrate that social intelligence is the ability to get along well with others and to get them to cooperate with you.

14 Introduction

References

Bauman, Z. (2000), *Liquid Modernity*. Cambridge: Polity
Castells, M. (2001), *The Internet Galaxy: Reflections on the Internet Business and Society*. Oxford: Oxford University Press
Cresswell, B. (1999), 'Embodiment, Power and the Politics of Mobility: the Case of Female Tramps and Hobos'. In *Transactions of the Institut of Brithish Geographer*, 24 (2), pp. 175-192
Deleuze, G. and Guattari, F. (1986), *Nomadology: the War Machine*. New York: Semiotext
Haraway, D. (1991), *Simians, Cyborgs, and Women: The Reinvention of Nature*. New York: Routledge
Levy, D. (2007), *Love + Sex with Robots: The Evolution of Human-Robot Relationships*. New York: Harper
O'Neill, J. (2004), *Five Bodies: Re-figuring Relationships*. London-Thousand Oaks-New Delhi:Sage
Ohmae, K. (1990), *The Borderless World*. London: Harper Collins
Prigogine, I., Stengers, I. (1979), *La Nouvelle Alliance. Metamorphose de la Science*. Paris: Gallimard
Prigogine, I. (1994), *Les Lois du Chaos*, Paris: Flammarion
Serres, M. (1995c), *Angels: a Modern Myth*. Paris: Flammarion
Turkle, S. (1984), *The Second Self*. New York: Simon and Schuster
—. (1995), *Life on the Screen*. New York: Simon and Schuster
Turner, B.S. (1992), *Regulating Bodies. Essays in Medical Sociology*. London: Routdledge
—. (1996), *The Body and Society*. (Second edition). London: Sage
Urry, J. (2000), *Sociology beyond Societies*. London: Routdledge
Webster, A. (2002), 'Innovative Health Technologies and the Social: Redefining Health, Medicine and the Body', in *Current Sociology*, Vol. 50, No. 3, May 2002, 443—457
Wellman, B. (ed.by) (1999), *Networks in the Global Village*, Boulder,CO:Westview Press
http://www.acm.org/ccp/references/wellman/wellman.html

PART ONE:

MOBILITY NODES AND SPATIAL MOBILITIES

THE SOCIAL BOND IN DIGITAL CONTEXT

PIERRE BOUVIER

Abstract: The social contract as we have known it for decades is today in disrepair. Major changes in social relationships have occurred due to the anorexis of the political apparatus and globalization. The 'corps social', that is the social 'body' or civil society is being transformed. Alternative forms of social interactions, fragile but determined, are being created from below. Endogenous social practices and symbolisms are re-emerging in postcolonial contexts both in thr Western world and in the former colonies. These constructs, groups and 'micro-bodies' are part and parcel of the interactions that emphasize physical, emotional and symbolic exchanges within society, rather than voluntarist preconditions and eschatology. These groups are obviously in interaction with the institutions that surround them and cut across them, but such 'micro-bodies' deconstruct and reshape them. Squats, endoreic practices, new forms of local groups are all postmodern constructs that need to be explained not only through the analysis of timeless social stuctures but also through the prism of their cultural practices and rites of belonging. They are shaping postmodern social bonds.

Keywords: social bond, political body, interactions, socioanthropology, collective practice, micro-bodies, postmodernity.

The question of the social bond and that of the body politic presents some points of convergence. Man is a political animal who can hardly disregard his fellow-creatures. Only hermits and anchorites deliberately choose to do without them. Recluses, prisoners or those who are alone in extreme conditions of incarceration may, against their free will and their wishes, belong to that part of humanity that has no or very few social bond to others.

The body politic, downstream, involves the notion of an ideal shaping of interpersonal relationships. This comes within the framework of a specific process aimed at initiating the condition of 'living together' – of an equally distributed 'body'. Participation in this body includes several levels, from the most motivating and involving to those requiring little or very individual input.

The social bond belongs to a much wider and more open dimension. While the body politic subsumes a reasoned and rationally organized intention, the social bond appears more 'immediate', in other words less mediatized by *a priori* assumptions, postulates and assertions – those of the body politic that frames it and points it in whatever direction.

This bond depends on the necessary exchange between human beings to both meet their survival needs and reproduce. The family is the main link in the chain. From the interplay between the latter and its human environments, various types of systems, both practical and symbolic, develop. These systems create numerous and varied social bonds, including the political bond, in the Aristotelian sense, with its focus on the city – that reason-based body of specific, regulating relationships. Social order and social peace postulate some organizing of the hurly-burly of human passions.

The social bond, therefore, seems more 'innocent' of any specific intent. It expresses itself with such diversity that the plasticity of bodies and minds usually prevails over regulated intentionality.

Its domains are multiple and its range of artifacts has the particular characteristic of being relatively free from the grip of immobility – that which, if not predetermined, is at least predictable and already in place. This plasticity of the social bond has the advantage that it can be a recourse in the face of systems and implications created by the body politic, in any given historical and spatial context.

The questioning raised by the concept of the citizen body, often defined as referring to the rights and the interpretative and active capacities of individuals, remains more or less dependent on a state of affairs that historically precedes it. The values present in this concept owe much to the antecedent and preconditions framing the possibilities of 'citizen' expression. Aside from the 'naturalness' that this citizen body acquires in democratic regimes – a much vaunted and recommended status – this set-up normally operates by means of institutional apparatuses, or at least in the space between the individual and organizations where an interactive relationship is played out. This relationship often takes on, quite logically, the traits and outgrowths of the various institutions that the state, the political parties or various officially, have legitimized. These vectors carry the capacity for establishing participation between and among the constituent parties.

As Machiavelli underscored, in his *Discourse on the First Ten Books of Titus Livius* (Machiavel, 1952), the organization of the Roman city, over a certain period, approached that balance, sought by the philosophers of antiquity, which blends monarchic, aristocratic and democratic regimes without falling into the errors of these three states: tyranny in the case of the first, oligarchy in that of the second, and demagogy in that of the third. The consuls exercised 'regal' authority, the Senate looked after the interests of the patricians, and the tribunes looked after those of the people. This representativeness was articulated around categories of social order, under the aegis of religion as the keystone and guarantee of the inescapable and apodictic truth of its assertions. It is what historiography has brought out, for example, concerning a tri-faceted people participating in the symbolic creation of medieval society: 'the medieval individual was caught in ̄a network of obedience, submission and solidarity (...) in general and for a long time these dependencies blended and prioritized themselves in such a way as to bind the individual even more tightly.' (Le Goff, 1982:258).The protests of the citizen body in the early years of the French revolution drew strength from the virulence of the opposition standing in its way. The 'Declaration of the Rights of Man and of the Citizen' (*La Déclaration des droits de l'homme et du citoyen*) of August 26, 1789, postulates in its Article VI: 'The law is the expression of the general will. Every citizen has the right to participate, either personally or through his representatives, towards its foundation. It must be the same for all, whether it protects or punishes. All citizens, being equal in the eyes of the law, are admissible to all dignities and to all public positions and occupations, according to their abilities and without other distinction than that of their virtues and their talents.' Several decades later, the radical declarations of 1789, and of their offspring, had become severely limited. Attempting to respond to this limitation, Charles Fourier presented, in 1808, a 'Theory of the four movements and of the general destinies' (Fourier, 1808). This proposition goes to the furthest extreme of a 'participative democracy', in summoning the polysemy of human harmonies in a theory of the primacy of bodies and passions. It determined a re-formalization of human relationships and social bonds, the breadth of whose spectrum and objectives is amazing even puzzling, to us today. In the prevailing conditions of the beginning of the third millennium, this type of reflection, like others of its ilk that marked the 19th century, can appear, for many of those official representatives or players in the instituted intellectual arena, as, at best, wild and dated imaginings.

The weakening or, at least, the dwindling of active and dynamic relationships – in the sense that they allowed expectations to be expressed and responded to – seems to be one of the characteristics of the passage from modernity to postmodernity. This withdrawal is accentued by doubt about the capacities of everyday democratic procedures to respond to participation. The civic body and the citizen debate may seem like an illusion, in which the orchestra conductor and his musicians are playing only to themselves – that is to say, to a public largely composed of audience-musicians only, repetitively playing virtually identical scores.

The sound is muffled to these bodies not belonging to the political orthodoxy.

Today numerous observations testify to the fragility of the social bond (Putman, 2000). This situation corresponds to various factors. Amongst the most significant, data relating to the economic sector and to social and political institutions stand out. The latter are struggling to give impetus to their own values: the prospect and, especially, the reality of a continuous and growing improvement in living conditions. As social conditions deteriorate, so does belief in the institutions.

Contemporary globalization is being accompanied, in parallel or by opposition, by a rise of 'local' values and practices. These are not only emerging in developed societies and those most affected by globalization but also in so-called peripheral contexts, in more radical and hybrid forms. Apart from the usual sociological data that can be classified according to recognized variables (social classes, institutions, developments, etc.) and studied by numerous schools, other evaluations have apperead, notably those coming from stance of post-modernism. New practices and representations are being formed which are a response to the growing alienation. For this reason, the explanatory systems – those of social progress and growing individual and collective mobility – no longer have the capacity to legitimize social relationships that so frequently fail to come up to expectations. Globalization, in its attempts at structural homogeneity, has also lost the ideological apparatus necessary to stimulate individual adhesion. The considerable pressure of the economic foundations is accompanied by only cautious, limited and inadequate legitimizations – and for good reason. Contemporary globalization has given free rein to the networks and connections of financial flows, with minimal interference. Economic liberalism is its main dogma and has become a sovereign body. Given this reality, the social fabric must reshape values sufficient to give meaning to peoples lives.

In the face of today's anomie, prior social contexts can be revived, either by those directly concerned or by institutions.

'Hot Endoréisme': by the actors themselves

Out of sync with or even hostile to contemporary values and the prospects of Western-style development, societies maintain certain social practices and ways of seeing the world. Supported by researchers and organizations keen to protect what is becoming increasingly marginal, a process of revival (rather than mourning) can be put in motion. Encouraged by these anthropologists and historians, the society opens its trunks, brings out its old tools and costumes, rehabilitates its memories and reconstitutes, in the present, the symbols and connotations that represented it yesterday and that will distinguish it again today. The (hot endoréique) of the words and traces is reincarnated. On a local level, it instills specific meaning – unusual, but relevant to those directly concerned, and to others.

Ethnologists, historians, socioanthropologists and curators talk about and exhibit 'Those over Here', just as 'Those Over There' are increasingly doing, in publications or in museums of traditional arts and culture. This process of restoring heritage to survivors makes it possible, to a degree, to climb out of the present and move towards an 'existing together', even if this is more imagined than real.

'Cold Endoréisme'

When there are no living witnesses, exogenous organizations can be charged with rediscovering latent meaning, (endoréique), hidden under the liana or amongst the rubble. These players – such as the British Museum or schools of Far-Eastern studies – decide to make ancient cultures or civilizations re-emerge and to restore their cohesion, despite the absence of living witnesses. By turning not to surviving human traces but to traces found by, in and on the stones, they compensate for this situation of lack and contribute to bringing out their significance. Although not directly linked to present-day questions, this significance still represents – by virtue of the themes or angles focused on by those discovering or reconstructing the semantic vestiges – a resource of some value in contemporary times. They construct meaning by playing consequences with past evidence. They put forward detailed scenarios concerning a place, an event or a culture. These operations of latent memorization recreate, rediscover or re-expose, in the framework of one or other individual or collective strategy, practices and values likely to bring about an 'existing together'. Historical people or historical facts are used to boost ones value, compensating for what lacks in the present.

These operations bring what is presumed to be rich in meaning up to date. Working with it is a self-distinguishing, self-honoring process. It's trying to lift oneself out of the 'unlisted' category (a dialectic process based on the little-known or the unknown) by using the relics of the past – bringing temples or pyramids out of oblivion. They must be dusted off and cleansed of the grime and excrement of the past, according to criteria supposed to be those of the initial model that also correspond to the expectations of the present. Heuristic works in the areas of the arts or politics are held up to view or exhibited at a distance, so as to find meaning where the immediate present-day is lacking or incompetent. It is a 're-existing together', if only intransitive.

In the framework of colonization, social bonding processes are established differently according to the context. One of the ideas in the British approach to expansion and the appropriation of territories reflected the fact that military and economic dominations and controls needed instituting first and foremost. Local administration – that of symbolic, religious and daily rules – was left to the discretion of the native populations, as long as these highly relative freedoms did not impede or go against the colonial domination. By authorizing certain autonomous areas and maintaining or even reinforcing certain local 'chiefs', this 'indirect rule' had the advantage of creating an alliance with some traditional representatives recognized by the native people.

In the case of the Australian continent, the process followed a very specific line. From the beginning of the occupation in 1788 until after WW2, the case for colonization reposed on the *terra nullius* concept. In the 18th century, an agreement with the other European powers hinged on an understanding that recognized the first to arrive on so-called 'virgin' territory (i.e. territory not already annexed by another European state or not showing signs of so-called 'high civilization' such as could be observed in China, India, etc.) had the right to consider it its own. This concept prevailed at the time of the seizure of the Australian continent by the British. For the colonizers, the various populations that did indeed occupy the territory were apparently few in number and scattered. We are led to suppose, from accounts by travelers who gradually, at the beginning of the 19th century, started going to explore this island-continent and, later, from the comments of colonists and administrators, that their way of life and customs were of the most rudimentary kind. Certain anthropologists, it seems, subscribe to the same view.

The native populations were dispossessed in the name of, and on the pretext of, at least an idea of emptiness, if not of true physical absence, and of a primitivism postulated by the first observers – to which the authorities clung. Since this might not have been enough, more specific factors were put forward, including a fundamental principle of European societies: the right of possession. It was first upheld that these were nomadic or semi-nomadic populations that had not defined any territory as their own. Apparently knowing nothing of livestock farming, agriculture and, *a fortiori*, industry, these 'primitives' had to be just as unsophisticated in matters of social or symbolic constructs. The absence of political régime, institutions and writings could, therefore, justify the installation of European migrants representing human progress. Neither the diversity of the Aboriginal groups (there were over 500 tribes) nor the multitude of languages and customs over a geographical area greater than Europe, weighed significantly in the face of the policy of immigration and monopolization and colonization of the space (Green, 2000). Very rapidly, in the space of a few generations, these colonists, came to consider themselves the legitimate inhabitants of this new nation they called Australia (Anderson, 1996). Henceforth, the term Australian would no longer apply (as in the 18th century) to the indigenous populations, but to the new migrants. The natives would now be called 'Aborigines' – a term more semantically civilized but one with generic and demeaning connotations.

The policies adopted were intended not only to play down the number of these aborigines but to actively reduce that number. Supposed natural and inexorable developments linked to progress and the advance of the human species would, most analysts thought, lead to their extinction. Representatives of central government went along with this trend. Successive generations developed a pastoral economy, dominated by the sheep – a space-greedy animal. The rapid progress of migration in the second half of the 19th century was accompanied by greater and greater land occupation. The aborigines being considered to have no specific rights, the fences enclosing the landscape drove tribes away or confined them. Exploitation of the subsoil brought the same result. The sacred and symbolic links that united the aborigines to the land, the plants and the animals were more than ignored. The introduction of sheep and cattle, agriculture and the drastic cutting of wood transformed entire regions, not only eliminating fauna (wallabies, kangaroos, emus, wombats, etc.) but also polluting sacred sites, rivers and billabongs (natural watering places). No longer able to survive as before from their close relationship with nature, by hunting, fishing and gathering, the aborigines had to accept the grouping conditions imposed on them by the authorities. They certainly became a readily available and manipulable labor force for breeders, mining companies and ever more numerous waves of European immigrants. In addition, diseases such as tuberculosis or leprosy, plus the ravages of alcoholism, contributed to a high death toll.

This dominant policy did, however, encounter some opposition. This came from various sectors of civil society, including ethnologists engaged in active fieldwork, continuing and confirming the ideas of A. de Quatrefages, who considered that the natives were not uncultivated nomads but hunting peoples, where: 'the family, the tribe and the nation were organized (…) and divided into genuine clans' (De Quatrefages, 1878: 334).

Ethnological studies took apart this supposed lack of sophistication. Between 1926 and 1929, W. L. Warner carried out an intensive study, entitled 'A Black Civilization: A Social Study of an Australian Tribe' (Warner, 1969), around the Murngin – a group of several tribes from Arnhem Land, in the northeastern part of the continent, near New Guinea.

This policy of elimination had a very long evolution, involving appropriation of lands and the evacuation of their inhabitants, who were grouped in awful conditions in temporary camps, in the heart of the most inhospitable regions. The marriages and half-caste children born of European and Aboriginal parents were pulled out of their family environment and shut away in camps run by religious orders, until the supervisory authorities considered that their primitiveness, not to say their savageness, was dissolved by contact with and guidance from the civilized world of the whites.

This remained official policy until the 1960s. As Barbara Glowczewski underlines in her book *Du rêve à la loi chez les aborigènes*, 'It is a paradox of colonization that the most ancient autochthonous peoples of the planet had to wait until 1967 for a referendum to grant federal citizenship to all, giving them, as well as the right to vote, the same rights as other Australians: to education, equal pay and social security.' (Glowczewski, 1991: 9)

This advance remains, however, subject to numerous contingencies, as can be seen today, in the towns and urban environments and even in the territories (for example in the North West) that the State has restored, albeit after long legal quibbling, to the Aborigines.

The site of Uluru (Ayers Rock) bears the marks. Already, at the beginning of the century, travellers had been impressed by the majesty of the Uluru and Kata Tjuta geogical phenomena. They became tourist sites: 'The tourist comes here, camera in hand, and takes photos of everything around him. What does he get? Another photo to take home that contains a part of Uluru. He should have another aim – to see right inside: then he'd no longer see the big rock. He'd see that Kuniya lives inside, as always. Perhaps, then, he'll throw his camera away (Tjamiwa).'(Kutitjulu, 2002)

The degradations caused by tourism, but even more current consideration of the inevitable 'Aborigine question', have changed the situation. The right to citizenship of the Anangus peoples is recognized, but is tightly framed within the context of the National Parks, to which the government grants a 99-year lease. Mixed management brings together institutional representatives and members of the Aborigine people. The latter try to influence the course of things, ensure their values are maintained and curb the curiosity and bad manners of the tourists by telling them about the symbolic nature of the place. It's not a theme park but an essential and living frame of reference and a site of rituals and *corroboree* that take place out of sight of the shuffling, gawking touristic presence. However, ever-expanding touristic demand and supply – involving hotels, restaurants, agencies and various services – attracts a non-Aborigine population coming from afar in search of work. There is a risk of an increasing disproportion between the hundreds of thousands of tourists (who, however, only stay for 24 hours, for the most part), these new immigrant workers and, on the other hand, the small number of autochtonous people.

Another Aborigine site, Gagadju (Kakadu) National Park, is vulnerable for relatively similar but specific reasons: it has been worn down by livestock farming, an army training ground and, even within its own area, mining industries. Here too, certain religious sites have been abandoned by the Aborigines and, in turn, by the ethnologists, before the flood tide of tourism – such as that for the Nourlangie caves and rock paintings. Secularized, the Rainbow Snake and the Namarrgon Lightning Man recieve only ignorant hommage. Once again, this will not bring about the definitive disappearance of these cultural billabongs and this ('endoréique') determination to preserve identities, without archaism.

Far from these protected sites, this lingering of living springs in the crevices of western society takes other forms, in the heart of towns. In the northern suburbs of Brisbane, on a site where a penal colony was established in 1824 (it was later moved due to resistance from the locals), something more than just traces is taking shape. An Aboriginal community trying to re-establish itself. It comes close to being, *a priori*, a 'practico-heuristic construct', in the sense that this attempt involves reconstituting (endoréismes) or re-emerging elements. It concerns the Kinpa Rings – sacred places, the plans of which had been drawn up before their destruction.

One essential point, howevern remains: the fact that these indigenous populations – the most ancient on the continent, dating back almost, 50,000 years – have not been exterminated, and that today they are even maintaining their social practices and rites. 'At adolescence, between fourteen and sixteen years of age, the children are taken into the bush. Combining painting, song and dance, the ceremony leads the uninitiated child towards a symbolic death. The seal of this journey is represented by circumcision. The child, to whom all the totems and symbols of his clan have been revealed and explained, becomes an adult and a guardian of the law. He will participate in life by living in this way, in a spiritual but real world. The metamorphosis accomplished, he returns to the rest of the community – the big family of guardians of the law.' (Serval and Godelier, 2002: 6). In this publication, Maurice Godelier adds some reflections on how hard it is for the the intruder to appreciate these pictorial ways of seeing the world. Indeed, certain types of awareness, while tending to folklorize these cultures through fashion, seize on music or the plastic arts to keep more political and radical claims at a distance. Assimilation will have to be subtler than before and proceed tactfully so as not to increase the partisans of territories under the total control and authority of the first occupants, as was the case only two centuries ago.

In 1992, the Mabo case (the name of an inhabitant of the Murray Islands – the most northerly of the continent, in the Torres Straits) came to an end, after over ten years of legal wrangling, with the High Court of Justice recognizing the proprietary rights of islanders having maintained a long-term relationship with their land. This was tantamount to a legitimate criterion for appropriation.

Extending this judgement to the whole of the country posed problems of quite another order. Resistance was considerable, coming from farmers, livestock farmers and industry, not to mention the owners of land split into lots. Up till now, in the balance of electoral power, the backbone of the parliamentary system, the elected representatives and the governments resulting from the polls have by-passed the problem by only recognising that the country's first inhabitants, the Aborigines, have a possible right to lands released by the State and partially leased. The leases are not title deeds but operating licences. It will henceforth be possible, if the original population can prove continuous presence, for these lands to be retroceded to them. Such advances are only partial and are up against the hostility of many. These include the sympathizers of the xenophobic and racist *One Nation* party, which considers the Aborigines too well treated – or the Alice Springs inhabitant for whom these decisions amountd to a 'waste of space' in favor of a work-shy population, living in camps, educating its children badly and spending the Thursday (pay-day) salary on alcohol. It is plainly hard for immigrant Europeans to conceive of a way of existence and a civilization that doesn't contain their own work norms, which are based on principles of material enrichment and profit. Numerous examples prove this, often from companies or organisations stoically reporting the absence of certain of their employees – who prefer a tribal reunion, a rite or a corroboree (ceremony), a hunting party or a fishing trip, etc. (these are almost 'bush time') to western employment norms or institutions. In Port Darwin, the organizers of a festival to present the Larrakia culture, in a museum framework, will take place, as an official from this nation indicates, in the fullmess of time: 'We've been discussing this for five years. It will be done in the coming months, and we'll do it according to our own rhythm – bush time' (Northern Territory News, 2002).

In the case of Australia, we have gone, *nolens volens*, from a colonial strategy of negation if not genocide to a post-colonial attitude which has to compromise with the populations that were able to resist the pressures of the dominant party.

Social bonding to the dominant culture has been a failure, considering the disparity between the number of Aborigines – several hundred thousands – and the material and symbolic capacities of the millions of European descendants, without counting the flow, albeit now reduced, of current migration. The eventual emergence of one or more nations of Aboriginal descent in Australia is unlikely. Contrary to what some political scientists would like to think, the course of history is never predictable.

The French Antilles

It is interesting to examine what has happened in another context, that of other islands – of the Caribbean – that have also fallen under European supervision. Doing so increases our awareness, by comparative sociological and anthropological study, of the different ways of treating questions of the social tie to culture.

As in Australia, there were populations living in this archipelago. In this case – that of the French Antilles – they did not share the ancient deep-rootedness of their Australian counterparts. They were mainly groups called Karib, originating from the South American continent. Fishermen and warriors, they went from island to island, considering them as theirs. Relatively small in number in the French Antilles (Martinique and Guadeloupe), they were much more numerous on the other islands.

The arrival of the Europeans, at the end of the 15th century, had dramatic repercussions. The Spanish expeditions, after Christopher Columbus, and the taking of possession of the Leeward Islands, including the future Haiti/Santo Domingo, Cuba, Puerto Rico and Jamaica, resulted in the radical decimation of the natives. In a few scores of years, only thousands were left, out of several millions. This was due as much to forced labor as it was to diseases or to the armed elimination of all those (and there were many) who tried to resist this colonization process. In Australia, the size of the continent and the fact that the Europeans only arrived at the end of the 18th century changed things only relatively. In fact, it was mainly the difficulty of reaching certain ethnic groups taking refuge in regions of little economic interest that, in part, preserved some of them. The Caribbean islands are small in size and the richness of their soil and subsoil easily accessible.

Although less rich in minerals, such as gold, The French Antilles nonetheless suffered the elimination of their first inhabitants. The occupation centered around cultivation, notably of sugar cane.

At first, the labor force used by the colonists to make these potential and actual resources grow was made up of natives. Once these were eliminated – and even before – the setting up of trading of slaves from the western side of the African continent presented itself as an alternative. The establishment and thriving of enslavement and trading of the African populations lasted from the 16th to the 19th centuries. Captured by native intermediaries or directly on the coasts of West Africa, these blacks were exchanged, sent to Goree, Oujda or the Congo and, after enduring exhausting journeys, sold and put to work.

In Australia, the putting of the Aborigines to work was effective but less extensive, given their limited numbers and the distances involved. In permanently occupying this island-continent and making it prosper, the English authorities relied on their own demographic resources and on labor from Europe. Racial 'mixing' only entered the colonization process in relatively recent years.

It was different in the Antilles. The Karibs only remained in certain particularly remote areas, as is still the case today with their survivors in the north-east of the island of Dominica, in a reserve tolerated by the central government.

Elsewhere, there were the Europeans and their descendants (relatively few in number) on one hand and, on the other hand, a majority composed of African slaves and their descendants. There was not and could not have been, *a priori*, an 'Aborigine problem' of the Australian kind.

All the categories, layers and classes originated elsewhere – on the one hand the whites, bringers of the values of economic, social and cultural domination, on the other the blacks, representing the opposite pole of poverty, lack and humiliation.

This co-presence brought about mixtures and the development of a cross-fertilization that, from the *beké* preserving his white roots, to the European Creole, to the Amerindian and to the mixed Black and White mulatto, runs the gamut of possibilities.

Social inclusion in or exclusion from the culture was inseparably linked to the colonizing process chosen by the supervising authorities. In the case of the French Antilles, the strategy adopted was one of assimilation. Secularism and the republican ideals of fraternity and equality allowed every level of the societies of Martinique and Guadeloupe to feel integrated in a common, shared project. Making the islands French territorial *départements* ensured the long-term attachment of these possessions to mainland France. The limits of this assimilation appear, however, in the extent to which some consider being colored to be an evil spell.. The idea that there is only hope for the dominant race is expressed in the novel *Je suis Martiniquaise* by authoress Mayotte Capécia who considers that she can only 'love a white, a blond-haired white with blue eyes…a Frenchman.' (Capécia, 1948: 59)

Conversely – or at least within a different strategy, given the points of departure of each context – for the Australians, the possible and expected disappearance of the 'native problem' (the problem of the original inhabitants) should eventually come about almost naturally. In fact, this will not be the case. Cultural inclusion is therefore a very live issue, all the more so now that an alternative culture, that of Aborigine social practices and symbolisms are re-emerging increasingly strongly.

In the Antilles, the problem of the autochthones having been resolved relatively quickly, for the reasons stated, the difficulty was to create good enough conditions to allow new arrivals, coming with the colonists, to share an identical temporal origin, at least on this plane, and be convinced of the reality of this cultural inclusion in the heart of the republican mould. While some, such as the author mentioned previously, and with her a significant number of Antilleans, do not fundamentally question this historical evolution, other writers, intellectuals, militant workers and union leaders are not content with this perspective. Aimé Césaire is one. In his *Cahier d'un retour au pays natal* (Césaire, 1983) he already makes clear in the title his full and complete belonging to the Antillean world and identity, implying an awareness of and rejection of the prevailing situation.

Césaire turned, for a while, towards the poetry and literature of negritude – a fundamental link or complement that allows him to re-identify the roots of the Antilles populations. Later, a political career based on the principles of emancipation integrated him into political dynamic of the mainland and its overseas territorial *départements*. This was not the case of the Frantz Fanon – a younger man than Césaire. Caught, at first, in the integration mould, he changed direction, due to contact with Europe, his medical studies, plus the fact that, in hospital contexts, he encountered immigrant populations from North Africa. *Peau noire, masques blancs*, (Fanon, 1952) his first book, published some years after *Cahier d'un retour au pays natal*, made him Césaire's counterpart. Taking the pretence of colonization and integration as a starting point, he focuses on the poverty and symbolic alienation that affects populations that are playthings for dominant white minorities and their economic and intellectual devices. This is the order of things on which the psychiatrist focuses his analysis. In this book, he presents a series of values and ways of seeing the world that weave a path and establish themselves between the categories of colonization. The living experience of the black Antillean touches many aspects of life. Fanon looks particularly at the question of language – of speaking the French of France and the vernacular Creole. Not without humor, he describes the pressure of integration into the dominant culture for Antilleans returning to Martinique after being in France: 'On disembarking, right from his first contact, the person sticks out: he replies only in French and doesn't understand Creole any more (…) After some months spent in France, a peasant returns to his family. Noticing a plowing implement, he asks his father, an old rustic type and no fool: 'What's that called?' In reply, his father drops it on his foot, and his amnesia disappears' (Fanon, 1952: 38).

Fanon also focuses attention on the relationship between the sexes and, in this context, between dominant whites and dominated blacks. Here too, he shows the effects of the crystallizing of perceptions to the advantage of the former and the disadvantage of the latter. In the conclusion to his study, the author, to the assessment of the situation given to himself and his fellows, refuses having to carry a historical weight linked to slavery, but also promises strong resistance to those who continue to reproduce an iniquitous order : 'If the white man disputes my human value, I'll show him, by making him feel all the weight of my humanity, that I'm not the stupid black sambo he persists in thinking I am.' (Fanon, 205-206)

Fanon made this prediction a reality in taking the side of the people fighting colonization and, more specifically, in becoming involved in the Algerian war of liberation. Concerning the Antilles, he remains skeptical as to whether they will be able to extract themselves from domination and from the latent integration into the parameters and values of mainland France. (Fanon, 1964)

Contexts similar to those of British and French colonialisms and post-colonialisms perpetuate, whatever options were chosen, the discrimination and domination suffered by the aborigines or the descendants of generations of slaves.

The Fragility of Social Bonds in Industrial Societies

Complex social bonds – those of work and everyday interactions outside of production and the concrete acts that they involve – suffer, like other bonds, from the effects of the economic crisis. The end of the 'Thirty Glorious Years' (1945-1975) denoted a growing weakening of the gains made by the world of work and of modernity as it was understood in the 20th century. Full employment meant salaries. Consumption was the order of the day and interactions related to the shared acquisition of goods and services played a significant role in the dynamics of major social bonds prevailing in industrialized societies in the years following WWII.

Today, unemployment or underemployment restrict this potential. The capacity of companies to create an endogenous bond is fading with the reduction of posts – both in number and specialty. Added to this there is regional delocalization or, more radically, the externalization of tasks beyond national borders. The personnel who both fed and were fed by social bonds constructed over time now have difficulty finding their place – even those who managed to escape being fired. A body such as a works council is finding its resources dwindling, notably in view of the reduced personnel. The benefits and social links it could create among workers and their families, thanks to leisure activities and cultural and intellectual exchanges, are decreasing. This weakening of the world of work and its attendant social bonds is causing entrepreneurs, knowingly or indirectly, to reappraise this type of mechanism – a constituent element of the social bond peculiar to industrial societies. Flexibility and even precariousness are tending to become the prospect for more and more sectors and a new discourse on flexibility hails its benefits while ignoring its pitfalls in terms of increasing reification. A series of fall-back structures are attempting to accompany this new postmodern order. The benefits that certain groups had created and imposed, including these mechanisms associated with the world of production, are struggling to keep and maintain the social bonds they had woven. In addition, they have to deal with the competition exercised, to their detriment, by the leisure-industries sector – which operates according to essentially commercial principles.

In the face of this anomie, individuals can choose to devote themselves to rituals of old habits: forms of sociability of slender relevance, that continue more or less to prevail through the messages and entertainments offered by the media. These desocialized or undersocialized individuals, insecure, find themselves drifting with the tide of incitements and preformatted programs. It seems premature, today, to anticipate the forms the social bond will take, other than by way of groups formed around specific, temporary interests – populational groups that may have meaning for some but which neither wish nor set out to tackle macro-social issues.

In the context of this postmodernity, the representative body, by virtue of the delegation it implies, is located, *a priori*, beyond the reach of individuals' activities. The latter express more often subcultures of specific problems. Various groups suffer from the effects of this reticence or at least of their limited interest in delegating the representation of their issues to a governmental or other exogenous groups. Individuals are currently more interested in creating forms of solidarity within the circles they frequent than in reiterating texts that no longer make sense. Rather, in the absence of prescribed librettos or credible and convincing scripts, they can be in a position through their dealings with each other, and the context of the emergence of a construct of specific and shared practices, to create for some and within their own subculture, new meaning. Should it prove sufficiently involving, this acquiring of sense may, eventually, lead to the constituting of a broader group. Dialogue via a common language can link and bind those who outside of, alongside of or even within the framework of institutions, can communicate with and recognize each other as belonging. Such groups may function (all things being relative) like ethnographic societies that produce their own practices and their own values. These groups are obviously in interaction with the physical and symbolic bodies that surround them and cut across them, but such micro-societies deconstruct and reshape them. In fact, that is what is actually happening, for example, in several cases arising from exclusion situations at the very heart of developed societies.

Inter-individual interactions are in play. They come into play without warning. Their dynamic is of low intensity. Their coming into contact grows from a starting point of implicit questionings. Usually, the exchange first concerns shared practices. Many of us have come across these practices, either by chance or in a more organized context. Festivities of regional origin, such as the Breton *fest-noz* (night festival), become, once transposed into a non-indigenous context, the means for some to rediscover meaning and express themselves alongside rather than in opposition to the means of production on which they depend. Such interactions have the capacity to reinstitute a clear but distanced tie to the immediate environment, be it physical, social or symbolic. These practices reproduced over a long period establish specific rituals that constitute a capacity to come together and unite. Perversions of procedures and programs that are normally and functionally used for the realization of some service or other take an unusual twist. Their purely rational and unequivocal aspects, as a simple range of functions, partially fade, as they take on a festive and playful role. These constructs of more or less illicit practices loosen restrictions. This strengthens and specifies the practices peculiar to the populational group concerned.

This experience comes more or less unexpectedly into contact with other experiences of the same type. So individuals who feel affected by the incomplete and ineffective nature of the body politic, turn to involvements on the fringes of everyday practices. This meeting with Others who are equally sensitive to the prevailing anomie and equally interested in ralliyng around similar ways of doing, brings about the shaping of a construct of practices that can produce and, indeed, will produce new meaning. These practical heuristic constructs are generated, in various forms, more spontaneously and naturally than proactively. Their dynamic is primarily endogenous. It takes little account of exogenous propositions, or only does so to the extent necessary for their existence. The intention, unlike those of highly ordered institutions, is not pedagogical. Neither are they concerned with proselytizing. Autonomy prevails, as does an irrational empathy. Fearing not so much the condemnation of instituted bodies as a lack of understanding of the motives behind their initiatives, some individuals come together to try to meet their own requirements directly. They distance themselves from conventional glosses and views, whose relevance no longer has the purpose it once had – that of a belief in a sharing of benefits under the aegis of a future defined by socialist or liberal ideologies.

The social imagination, as Cornelius Castoriadis points out, exceeds institutional determinism: 'The self-alienation or heteronomy of society is not 'simple representation' or the inability of society to be represented otherwise than as instituted, from afar and at a distance. It is incarnated and strongly and heavily materialized in the concrete institution of society, incorporated in its conflictual division, carried and mediatized by its entire organization and endlessly reproduced in and by social functioning – the 'this-is-how-it-is' quality of objects, activities and social individuals.' (Castoriadis, 1975: 497-498).

Eventually, these constructs of new practices may – if their characteristics concerning the producing of meaning go beyond the circle of the initial players – give rise to the desire to continue. Potentially, these constructs can unite expectations and find further resonances. They occur in several forms. They form a body and can, because of this, according to certain conditions (in particular that of relevance going beyond the framework of endogoneity), open up to 'coherent populational groups''of a broader scope. Against all rational logic or, on the contrary, in a latent and un-clarified fashion, this tendency can allow the updating of ways in which the social bond can be reconstructed – from micro-bodies to actions. This is the case, for example, of artists (young or otherwise) for whom finding even physical space conducive to their desire to express themselves is something of a challenge. The absence (or very limited number) of artists'studios available inhibits their possibilities of exercising their talent. Some of them band together, without the backing of the authorities, to take possession of unoccupied buildings. They do so discreetly – at least at first. Later, despite the risks, they may decide to open up to the outside. In so doing, they create a social bond which goes beyond the initial construct and the endogenous context. Populational groups can then emerge, claiming – as organic, intellectual, plural, political bodies – to be founded on a coherence born of unusual practices and perceptions: those of the will to express themselves personally but within a collective framework.

In this context, the issue and debates concerning democracy, whether designated as participative or representative, seem (like the notion of citizenship) to be lest important than an understanding of the social bonds within a democracy. Tensions and disagreements can be resolved by the springing-up of new shoots – new constructs potentially capable of producing populational groups that can carry the social bond to where changing currents meet, on both sociological and existential levels (Bouvier, 2000). The tendencies are more towards self-dissolution, evanescence and transience than towards the constituting of institutionalized bodies.

Street parties, neighborhood associations, the defense of some temporary or limited issues, collective mobilizations, etc., show this modesty and this desire to do something oneself, without an intermediary. These constructs, groups and micro-bodies are part and parcel of the interactions that emphasize physical, emotional and symbolic exchanges among micro-bodies within society, rather than voluntarist preconditions and eschatology (Bouvier, 2005). Such attitudes may suggest, in the mid-term, the possibility of strenthening ties and 'existing together' in ways that the collective postmodernity has yet to acknowledge.

References

Anderson, B. (1996) *L'imaginaire national*, Paris : la Découverte.

Bonniol, J. C. (1992) *La couleur comme maléfice*, Paris : Albin Michel.

Bouvier, P. (1971) *Fanon*, Paris : Ed Universitaires.

—. (2000) *La socio-anthropologie*, Paris : Armand Colin.

—. (2005) *Le lien social*, Paris : Gallimard.

Capécia, M. (1948) *Je suis martiniquaise*, Paris : Corréa.

Castoriadis, C. (1975) *L'institution imaginaire de la société*, Paris : Seuil.

Césaire, A. (1983) *Cahier d'un retour au pays natal*, Paris : Présence africaine.

De Quatrefages, A. (1878) *L'espèce humaine*, Paris : Germer Baillière.

Fanon, F. (1952) *Peau noire, masques blancs*, Paris : Seuil.

—. (1964) *Pour la révolution africaine, écrits politiques*, Paris : Maspéro.

Fourier, C. (1967) *Théorie des quatre mouvements et des destinées générales*, Paris : Jean-Jacques Pauvert.

Glowczewski, B. (1991) *Du rêve à la loi chez les aborigènes* : Paris, PUF.

Green, N.L. (2002) *Repenser les migrations*, Paris : PUF.

Jorion, J. and De Meur, G (1980) ' La question Murngin, un artefact de la littérature anthropologique ' *L'Homme*, t. XX, n°2.

Kupka, K, and Testart, A. (1980)' A propos du problème Murngin : le système des sous-sections ', *L'Homme*, t. XX, n° 2.

Le Goff, J. (1982) *La civilisation de l'occident médiéval*, Paris : Flammarion.

Machiavel, N. (1952) *Œuvres complètes* : Paris, Gallimard.

Mutitjulu Community and Anca, (2002) ' Bienvenue au parc national de Uluru Kata Tjuta '

Northern Territory News (2002).

Putnam, R. D. (2000) *Bowling Alone*, New York : Simon & Schuster.

Serval, A, and Godelier, M. (2002), *Wati, Les hommes de loi*, Paris : Woo Mang.

Varembey, A. (1889) ' Australie ', *Dictionnaire des sciences anthropologiques*, Paris : Doin/Marpon et Flammarion.

Warner, W. L.(1969) *A Black Civilization. A Social Study of an Australian Tribe*, Gloucester : Peter Smith.

STABILITY AND WANDERING: SELF, COHERENCE AND EMBODIMENT AT THE END OF THE SOCIAL

JEFF VASS

Abstract: The question raised for this book might be posed as 'what are we *present to*, or how are we *present to ourselves* within our own activities'? The mobility of the tourist as a response to the difficulties of *presence* in modern life has long been associated with the concept of the *gyrovagus* in Benedictine thought: the wanderer with no *stabilitas*. The metaphor of the tourist constitutes for us an image of how we are now situated within our own activities and there is a fear that ICTs form a new medium for the 'gyrovague' to wander. ICTs are a source, and a resource, of what Habermas (1987) and Beck (1992, 2006) call 'mediatisation' a process which penetrates and transforms 'presence' and, for Habermas, takes us further from the 'sacred basis' of social life. This paper argues that the problem of contemporary life may be better grasped theoretically as a *sensory* problem of coherence within limits. The new importance of 'bare' or 'naked' life and 'remainders' (cf. Agamben, 2002 and Foucault, 1994) has shown us the difficulty we now have with what limits construct. However, this paper develops our understanding of this difficulty as a problem of 'subject-body coherence in zones of social making'. Remainders are always a product of social making. It is further argued that mobility and stability are always with us as features of 'zones of social making' but, in their new conditions, represent new problems in dealing with constructing presence, its limits and coherence.

The sociologies of mobility and flow (e.g. Lash, 2002; Urry, 2002) have arisen against the background of some lack of confidence that the idea of 'the social' has a future (Gane, 2004). As Baudrillard (1983) prophesied it may be that we have now arrived at the 'end of the social' in any case. These perceptions of the fading of the social world, as traditionally conceived, appear to be corroborated by the ways in which information and computer technologies (ICTs) have emerged and have most recently transformed global networks and the domestic, leisure and work spaces we inhabit. These networks were once imaginable, even ten years ago, as having a 'social' origin and objective, for example, in the form of network imagined by Castells (1996) in *The Network Society*. For Castells the 'spin' on the loss of traditional social connectivity is positive. In the positive view ICTs *assist* a 'new world order' enabling us to imagine lives no longer as multi-layered 'life-*courses*' constrained by lack of spatial and social mobility but as mobile, orchestrated and intersecting work, leisure, learning and life *sequences* (Vass, 2002). The consequence of imagining our activities and bodily involvement with the world as globally extended and diversified *sequences* enables us to at least imagine global redistributions of resources and skills, even at the price of some loss of coherence of the 'self'. In the network regime the self becomes a project in its own right. Written again today, however, for reasons to be explored further here, Castells' magnum opus might be called simply *The Network*. For Baudrillard, however, the end of the social carries a negative 'spin' as this end was inaugurated by, what he sees, as the peculiar character of our inhabitation of a hellish 'mediascape'. For him the same extended bodily involvements lead to an absorption of anything that was once social into a nightmarish 'hyper-reality'.

Between these positive (networked *utopia*) and negative (mediascaped *inferno*) perceptions lay the sociologies of mobility and flow. At one and the same time they appear to both celebrate the 'liquification' of modernity (Bauman, 2002; Lash, 1999) and fear its consequences. This mixture of celebration and fear is symptomatic of a sociology that is excited by the way it has come to read the novel circumstances that appear to embed us. However, I shall argue that this mixture of positive and negative reactions which we witness in the flow and mobility literature is predicated on quite ancient, and specifically European, categories of thought: human stability and wandering. Furthermore, I shall suggest that the body and its labour, conceived as an 'activity field', has become the principal arena in which are played out the traditional anxieties and possibilities enshrined by stability and wandering.

Bodies lack wholly adequate theorisation in this context. Again, to be crude, approaches tend to be one of two kinds. Either critical appraisals of the body as a 'constituted', 'produced' or 'represented' entity (e.g. Howson, 2004), or they promote a micro-sociology of engagement and involvement with the world, often inspired by the work of Merleau-Ponty and Bourdieu (e.g. Crossley,2001a). Recently, there has been some controversy over whether the latter kind of work, centred on an analysis of the 'lived body', made famous by phenomenological approaches and taken up by Bourdieu, are even sociological at all (Howson and Inglis, 2001 ; Crossley, 2001b). Howson and Inglis (2001) argue that attempts to understand the body *in its active involvements* fail to make the critical connection with aspects of social structure which (they say) are the principal concern of sociology. Yet, the concepts of flow and mobility, which attempt to move beyond traditional conceptualisations of social structure, gain their very momentum from what can be said of the interface between agency and structure and in those active involvements even if the idea of flow empties the traditional terms of agency and structure of some meaning. This suggests, at least, that there is room for some theoretical work here to try to understand how the body, or rather embodiment, is implicated in the kinds of flow under review.

Outline of the issues

While ICTs, such as the Internet, may be seen as fostering flow and the creation of new networks in the form imagined by Maffesoli (1996), for example, where new groups are created through mutual concerns (single issue politics; online friendship and dating groups; consumer interest groups etc.) the benefits seem to come at the loss, or in the wake of the loss, of traditional sociality and what we take to be the grounding conditions of social and personal stability. In short, the Internet and mobile telephony (including its mobile Internet, imaging, data-storing and time-organisational capacities) *appear* to set up for us entirely new 'portals' through which the dynamics of sociality, stability, wandering, anomie and alienation, or their replacements, need to be rediscovered again. The notion of 'portal' here is somewhat misleading. In the context of the language of Euclidean space it suggests already-embodied subjects moving across an already constituted territory, and is of less certain application in the topology of the Internet. Interface and boundary have the same problem. Part of what needs to be addressed is not an *ergonomic* relation between bodies and ICTs but a sense of a subject-body-ICT field along the lines imagined by Donna Harraway. However, I am interested not only in how this field opens up new territories in which to wander, but how it rewrites traditional human concerns about stability and 'being on the outside' – as I have suggested before 'a tourist in one's own transactions' (Vass, 1996). In particular, ICTs seem to belong to a category of media forms that impact on the nature of co-present subjects and the 'lifeworld' that embeds them. It is worth rethinking these two things together.

To begin this task and move forward, for me, means going backward in three ways.

1. The first departure is to try to examine how and where stability and wandering show up as issues through the links made by human groups between their circumstances, their understanding of stability and the dangers of wandering. This provides us with a map of stability and wandering in the contxt of historical changes to the experience of loss and incoherence.

2. The second departure is to look at those sociologies that warn that the powers of the 'social' to ground stability and coherence are *fading*. Here I ask what the nature of that ground is and how is it threatened by forms of mediatisation or replaced by flow. As my examples are drawn from communication my focus will be on Habermas and Lash. In both cases it will be seen that there are unresolved issues of embodiment at the core of these ideas.

3. It is the issues surrounding embodiment that will permit us to reconnect current concerns about ICTs to chronic human themes.

Stability and Wandering as Chronic Human Themes: Prelude to the Sociologies of Mobility and Flow

Stability and wandering have been key themes in European and Middle Eastern cultures since at least the earliest known oral literatures. I would contend that contemporary sociology itself forms a literature with much closer ties to ancient literary output, through its concerns with exactly the same key themes, than sociologists would like to believe perhaps. By 'key theme' I mean that human groups have paid attention in profound and concerned ways to aspects of their social, cultural and embodied experiences that appear to them to connect with *their grasp of their circumstances* in all their tangibility. The written versions of Homer's *Odyssey* and the Babylonian *Epic of Gilgamesh* may be thought of as Iron Age redactions of much earlier oral memories of the Mycenaean and Sumerian Bronze Ages respectively. In both ages we are dealing with, very different, social formations (cf. Giddens, 1981) that are concerned to rehearse the nature of what constitutes stable relations and what happens when we wander. As well as danger for the social group, wandering implies a loss of the stability to which we moderns give the name 'self'. Gilgamesh, as king, of the city-state wanders dangerously away driven by urges impossible to satisfy: desire and the search for immortality. While on his adventures he 'finds himself' and re-establishes coherence by accepting that death is inevitable. He is then able to 'return'. The story of Odysseus's return is well-known. He leaves his city-state, Ithaca, with the consequence that it becomes a de-stabilised community. Oddysseus's wanderings become partly a struggle against the dissipation of self-hood from, literally, being blown around by forces beyond his control. His return to Ithaca culminates in having to recover his 'true self' through setting to rights the social relations that have become chaotic in his absence. The self, made coherent again, is a co-terminus event with 'right social relations' re-established. Place and coherence are deeply connected when wandering is constituted by spatial mobility in a geographical sense. And in these examples geography is to be taken in a territorial sense. Both Gilgamesh and Odysseus are rooted spatially in their *territories*. The stabilities that emanate from their selves and through their households into their communities are symbolised by a foundation stone in the story of Gilgamesh. A rooted tree forms part of the structure of Odysseus' house, and indeed, his marriage bed. Kinship, status, authority and their stabilities within place defined by territory could not be more clearly imagined.

Christianity de-territorialised the relation I have proposed here between place and coherence. The social and moral world of the first Christians (Meeks, 1999) was diasporic and heterogeneous. It was spread around the multi-ethnic Greek-speaking Hellenistic world. Although the young Church, as a 'glocalised' organisation, strove for stability strategically in the codification of its public expression of creeds, regulation of its liturgies and sacred texts, we can still turn to its favourite *stories* to get a flavour of the de-territorialised coherence that was of significance to its members. That is to say, we can sense the way that stability began to 'show up' for these communities. New Testament stories and parables such as that of 'the prodigal son' (Luke 15:11–32) contrast *coherence*, as 'path' (to God), with *wandering* as a veering off from the path leading to dissipation of the self and, importantly, *waste*. Wandering is a kind of 'squandering' of self. Off the path the self is 'migrant', placeless and squandered. It becomes a 'remainder' in the sense explored by Agamben (1998) in *Homo Sacer*. On the return of the prodigal son the father reflects, 'what was lost is now found again' (Luke, ibid.). The development of Christian thought in the West gave rise to two very distinct senses of the loss of self. On the one hand, loss of self *on* the path was a high Christian virtue and promoted within, for example, Benedictine monasticism between the fifth century and today. The meaning here is to 'lose oneself' in the project of salvation shared by the community. On the other hand, loss of self, as incoherence and dissipation, is associated with being *off* the path by being a consequence of movement away from it. This is sin. In orthodox Christianity sin is still etymologically associated with erring and wandering: *plane* in Greek means wandering, giving rise to the word *planet* which means, of course, 'wandering star'. The body may have become the site for the management and control of wandering originated by desire through asceticism and penitential activities (cf. Tambling, 1990), but that is not my primary focus here. Rather, in the first instance, it is the form of coherence of the self that *shows up* in opposition to wastefulness. This form of coherence is characterised by 'being present to oneself in the here and now' and is not territorial. Indeed, for Christians, it is important that this form of coherence can occur while 'detached' from material things and places. It is formalised, for example, in the Benedictine tradition of the practice of 'recollection'. After the labours of the day it is often the practice to have a period of recollection.

I shall return to Christian theme of de-territorialised stability and wandering later when considering mobility and flow. For now, it is necessary to note that the key themes of stability and wandering continue through the traditions encountered during the Christianisation of Europe between the fifth and fourteenth centuries of the common era. The Icelandic Sagas, for example, which became written down and Christianised during this time were themselves derivative of oral memories of continental migration and separation from home. Hence the tales themselves *show concern* for the dynamics of stability and wandering, their consequences for selfhood and the stability of social groups. Communities formed as a consequence of migration, and who were aware of the threats posed by the migration of others became over-sensitive to the status of the wanderer (Byock, 1990). Wanderers are social *remainders* who are actively excluded from being in the group, co-present with others, and therefore the means to be present as a coherent self with a fully human identity. This includes the community's production of wanderers by their expulsion from the group. In *The Saga of the Volsungs* (incidentally, one of Wagner's key texts for the Ring cycle) the hero's father Sigmund is driven from society and away from other humans, and with a second son, they live in an underground cave, and 'dress in wolfskins and howl like wolves' (Byock, 1990). Again, one of Agamben's Durkheimian themes is paralleled here. The ability to create or define social remainders, the power to exclude from 'co-presence', is one of the defining features of sovereign power. Agamben (1998) finds that the Christianised Roman law developed in Europe constituted a spatial distinction between civic society and the area outside the city: a wasteland whose inhabitants were *vaga*bonds, wanderers. To ban someone, as Agemben points out, quoting Calvalca, has serious consequences: 'Whoever is banned from this city on pain of death must be considered as dead.' ...Germanic and Anglo-Saxon sources underline the bandit's liminal status by defining him as a wolf-man [...] The life of the bandit, like that of the sacred man, is not a piece of animal nature without any relation to law and the city. It is rather a threshold of indistinction and of passage between animal and man, *physis* and *nomos*, exclusion and inclusion.' (Agamben, 1998: 105)

The 'threshold of indistinction' between nature and law here means that man (sic) so excluded enters a zone where s/he *belongs to neither*. But this indistinct threshold also holds as a boundary more generally between coherent presence and the incoherence of wandering. Agamben's concept of 'indistinct thresholds' I feel is an important one in the attempt to think about the new kinds of stability and wandering posed to us by bodies in the 'technosphere' (cf. Cregan, 2006) where the resources of the technosphere, ICTs, are formative of, or are in collusion with, the mobilities and flows which have become the concern of sociology. There is an important distinction to be made between threshold of indistinction and the kinds of boundary explored by Mary Douglas and Judith Butler in relation to the body. I will return to this.

Here I seek to focus on the coherence, the sense of loss and dissipation that its disappearance portends, and around which there is both fear and a striving. This is always of perennial concern to people. It is not merely a contemporary concern. I am not arguing that key themes in early epics are in some way 'proto-theories' of an ontological construct which we call the social. I am interested in the character of the concerns human groups have that arise from, and gesture towards, their experience of the circumstances that they construct and embeds them – that is circumstances which show up in their concerns and their attention to cultural forms that touch on them. These concerns and their reflections, through whatever media are available, are related to what Lyng (2004) calls a group's mode of 'bodying forth'. In more literary forms of expression we might refer to a group's strivings.

I have claimed already that contemporary sociology is a 'medium of rehearsal of concern' with stability and wandering. This is shown explicitly in the mobility turn. In this regard sociology is like its epic forebears. However, unlike epic, drama and poetry sociology is considering the issues at a point in time when its founding concept, 'the social', is under attack. The concepts of mobility and flow appear to some (e.g. Lash, 2002; Urry, 2002, Beck, 2000) to overtake the idea of the social. The future of our sociological understanding of stability, wandering, the coherence of the self and its modes of loss will depend on how we generate concernfulness and about what. In the context of the body-ICT field we need to be aware of how human strivings are engaged within the technosphere, the new 'theatre' of stability and wandering. Technology promotes mobility and flow and in this regard can be said to aid and abet the processes which threaten the social. Information and computing technologies relate to the problem of the end of the social in at least two ways. Firstly, they can be seen as a force complicit with modernity that produces a *reduction* of the social and a *minimisation* of the lifeworld as conceived, for example, by Habermas (1987), Giddens (1981, 1990) and Bauman (1993, 2000). The reduction and minimisation of the social bears upon the fate of co-presence which, historically as suggested above, has always been a key element of the theme of stability and wandering. The body-ICT complex in this first version of the end of the social is a site of what Habermas and Beck have referred to as 'mediatisation'. The mediatised self requires new strategies for coherence and comes to rely on resources which are not immediately co-present. Secondly, the ICT-body complex features in the concept of flow. Here the end of the social refers to the idea that flow effectively *absorbs* the social into a circulation of objects and information. The issue of absorption poses different kinds of problems to the possibilities and strategies for the coherence of the self as well as its dissipation and loss. I will next explore these two versions of the end of the social and suggest how they relate to the question of the body-ICT field.

The 'Fading Social': Mediatisation and the Problem of Coherence

Coherence of the self and the key themes of stability, wandering and loss came to be swallowed by the semantics of 'the social' in the founding period of sociology. As Dowling (2001) explains, during this period the idea of the social became 'ontologised' in the work of Durkheim. The theme of stability disappeared into the shadow of the concept of 'integration'. Loss of coherence and self might be aligned with the concepts of anomie and alienation. Selfhood and its coherence have always been supplementary to the question of society in the twentieth century, except perhaps within the somewhat niche approach of phenomenology. It is only more recently, in the wake of the fragmentation of the social that the shadows cast by monolithic social theory have parted to reveal the key themes once more. The first type of argument about the fading of the social I consider here emerges from thinkers like Habermas and Giddens who remain wedded to the ontological enterprise and see recent social transformations in the light of a falling away from forms of integration. However, the focus of their attention is very much on the experience and conditions of co-presence that sustain social life. My argument here is that beneath their heavily 'cognitive' focus on social interaction and communication, which are seen as disrupted by forces that dislocate the place of presence from social systems, lay an unresolved problem of the body and its social connectedness. In both cases their understanding of the minimisation of the social takes us to a narrow interface between co-present selves and the global systems that seem to be transforming us. Yet, digging deeper, there is in the sources of their material an opportunity missed to focus on the question of embodiment. The body and its media of action become the site of stability, wandering and loss of selfhood.

Arguments about the fading or minimisation of the social do two things. Firstly, at societal and global levels they examine the intensity and extensity of transformations to the social, political and economic system connections of high modernity (e.g. Giddens, 1981; Habermas, 1987; Held *et al*, 2002). Secondly, there is a concern with the impact of these transformations on fundamental social experience. Giddens' work in the 1980s and 1990s was preoccupied, for example, with the fate of human co-presence and the living encounter. Similarly, Habermas (1987) in the development of his theory of communicative action, was concerned with the 'structural violence' that transformations to the system wreaked on the domain in which actors live out their lives, the 'lifeworld'. For Giddens and Habermas the minimisation of co-present social experience produced new *qualities* of human transaction brought about by *mediatised* institutions on the lives that support them. These theoretical perspectives were developed largely before the advent of the mobile phone and the Internet.

To focus on co-presence and the living encounter in these ways is to ask questions about the 'moral bindingness' (Giddens, 1981) of human life. Indeed, in Habermas (1987), Giddens (1981, 1990) and Bauman (1993) there is a strong sense that the conditions of high modernity have *penetrated* in dangerous ways the very ground of social life. This is a threat that impacts on co-presence and in many ways is the resource and product of moral bindingness. All three thinkers are concerned to examine a dislocation between the lifeworld, a domain of action, performance, involvement and experience, and the 'system' a domain of structural connections increasingly globalised and transformed by technology. Part of the problem with the approach, however, has been the non-compatibility of the languages of lifeworld and system. Thus Giddens, for example, draws on phenomenological metaphors to deal with the experiences of high modernity such as 'ontological insecurity', while using theoretical terms from geography to trace the new dynamics of the globalised system. Furthermore, there is not a detailed elaboration of the bodily underpinnings that their analyses of the fate of co-presence implies. Moral bindingness for all these writers is a drama that plays out in the domain of communication at a cognitive level.

For Habermas and Giddens I suggest that co-presence and the positive meanings it has for selfhood, group and coherence is primordially a concept of *embodiment* and bodily engagement. Either this is hidden from them, or they have chosen to ignore it in pursuit of projects with a cognitive bent. Unlike for Bourdieu, the body issue did not show up for them in the same way. To explore the impact of mediatisation in Habermas and Giddens they are inclined to shift our attention away from the responsiveness, engagement and activity of the body, which I now will claim lies at the root of the problem of co-presence and moral bindingness in their own work. Instead, they take us directly towards *social interaction,* conceived as a communicative event in the linguistic sense, where any further impact of the forces of modernity become problems of *interface* between actors and media. An example of this would be the way in which Thompson (1995) extrapolates this discussion of modernity in the context of the media. Thompson (1995) formulated the problem of mediatisation as an issue of communication and interaction in alignment with Giddens and Habermas. He identifies three types of interaction: face-to-face; mediated and mediated quasi-interaction. Thus, co-presence becomes a 'context' for face-to-face interaction where bodily movements, winks and gestures are *supplements* to the main business of the exchange of messages. Mediated interaction stretches across space and time by virtue of the use of telephone, writing and other technology and it is seen to differ from face-to-face largely as a consequence of spatial distinctions between the interactants. The *problems* posed by these media require interactants to orient strategically to the media in order to get their messages across because they lose the supplements of communication. Mediated quasi-interaction involves the actors involvement with institutional media. This provides a one-way monologic flow of information towards no specific actors.

This schema suggests that Internet media communication merely constitute an interface through which system messages travel to sites of social interaction. However, it does not capture, for example, the field of involvement that opens up in the world of on-line communication. Internet dating sites or 'fan' networks, for example, are often multi-textual constructions in which all three types of interaction defined above co-exist. On one screen is a web-cam connecting two users or an IM instant messaging system which allows moment to moment text-based response communication, a 'profile' of the interlocutor which may be addressed to all users of the site, generalised rules of engagement provided from the site itself, pictures of the interlocutor taken at different times and places and lists of their likes, dislikes, expectations and so on. In many ways it can be argued that the subject -body-ICT field produced here problematises any simple notion of self, co-presence and context.   Furthermore, in any developing on-line relationship there will be consequences for the non-electronic locations occupied by subject-bodies.   The problem of co-presence here is one where the media involved enter into *sensory connections* with the bodies they link.   The sensory modalities of the body-ICT field begin to manage the *responsiveness* that is a vital pre-condition of co-presencing as well as the more hermeneutical textual material that is brought into new and unusual relationships with it. These new relationships are predicated on how the body, whose indeterminacy was disciplined in a social relation, is now managed in a field of ICT-organised contingencies. It is the sensory challenge posed, in this instance by ICT, that forms the same origin point of Habermas's sense of the social.   However, he misrecognises it.   I now turn to this to trace Habermas's thought back to the embodied conditions of pre-cognitive sociality.

Habermas wants us to consider the question of the social as communicative activity in performance in the lifeworld. The condition of modernity is that the lifeworld is 'uncoupled' from the social system. In order to appreciate what are now the special conditions, dangers and difficulties pertaining to this 'uncoupling', we need to understand how Habermas construes 'the social'.   For him the very foundations of the social are the object of Mead's attention.   Though Habermas over-emphasises Mead's end point. That is to say, the moral commitments and the ethical contours of social life are rooted in the norms that situated actors communicate in their expressed expectations and obligations, in other words the *mutuality* that enables socialisation and co-present communication. However, the basis of social life lies, according to Mead (1934), in reciprocal interpersonal bonds *that come to have* a rational basis in the communicative practices of the actors involved. This sounds then like this should be developed as a 'cognitive' project but should not be only a cognitive project. A *closer reading* of Habermas's discussion (1987, Vol 2) shows that Mead, like Habermas, is concerned with the *character and quality* of the 'social bond' in, for example, what constitutes *sincerity of engagement in conditions of bodily responsivity.*

Habermas's next move, after jumping from embodied responsivity to a *second*, linguistic basis of the social, is to understand the shift from the 'sacred' character of the bond to the contractual one that depends on an already established language game. The Durkheimian problem here is familiar: what kind of contract establishes what contracts bind? At this point Habermas reverts to his Meadian principles and emphasises that people bound in a contractual arrangement have a sensuous relation to it. Their understanding is of a *sensorial* and not a *cognitive* form. He says people need a publicly maintained 'story' *through which the parties can 'sense' their statuses and the nature of the contract* (ibid.:80 ff.). He does not say exactly in what this 'sensing' consists. But it is important to note that the 'story' might be thought of a sensuous instrument *before* it has a cognitive status open to rational exploration in discourse. Discourse then pulls itself up by its own bootstraps,

> To the extent language becomes established as the principle of sociation, the conditions of socialisation converge with the conditions of communicatively produced intersubjectivity. At the same time, the authority of the sacred is converted over to the binding force of normative validity claims that can be redeemed only in discourse. (Habermas, 1987 Vol 2: 93).

So, through Mead he develops the basic assumptions of a 'communicative ethics' based on the 'ground' of intersubjectivity. But we should note a significant jump from the *sensory and embodied qualities* of intersubjectivity and sociation derived from his reading of Mead to a communicative ethics which construes language not as a sensory instrument anymore but as a servant of rationality. Ironically, it is rationality that ultimately implies 'delinguistified' media at system level (administrative mechanism, money etc.) which penetrates co-presence but which is now *disconnected* from norms and values. Activity in the lifeworld is *marginalised and remaindered* in relation to these mediatised 'monolithic projections'. Administrative mechanisms, repertoires of bodily procedures and money, as objects in the body-activity field are not now considered, or returned to, in their sensory qualities.

Loss and incoherence of self become features of the lifeworld and performativity in this model. It implies that the use of rules and norms becomes a hermeneutical rather than an embodied activity. Sadly, Habermas tells us nothing about this dimension but everywhere seems to rely on it. If co-presence becomes disconnected from norms and values, the real issue is not that such things have disappeared, but that they have ceased to be *embodied* and *sensorial* instruments of exploring the contingencies of the lifeworld but rather 'texts' that give rise to hermeneutical experiences. The lifeworld loses the *stabilitas* offered by sensorial familiarity. As Bauman (1993: 119) says discussing the same issue 'I must trust the masks'. In this sense Internet sociality poses additional rather than different problem for self coherence. There are masks on *both sides* of the computer screen. In this scenario we see the recruitment of selves into an ICT domain to which they do not belong. At the same time the domain constitutes the place in which the body is as a place of no belonging   as well (i.e. as a mediatised space)! The transformed conditions of body-field contingency and multi-textual encounter posed here raises questions. Rather than an ICT interface as a social-media boundary separating interactants, might we not speak of a 'threshold of indistiction'? To recall the Agamben theme again at this point, we may be able to point to the lifeworld and the media systems it supports as a fading domain of sociality and stability. But in the context of the new ICT media a new form of wandering begins to be produced. This kind of wandering has no *from*, only a *to*. The only stability that remains is the indistinction of a body-ICT field itself.

The Social 'Absorbed': Flow, Information and Embodiment

It is, perhaps, in the sociology of flows, where the social is absorbed by culture and its vicissitudes, that the human story of stability, wandering and coherence reaches its most unpredictable turn. What for Baudrillard (1983) is a nightmare of hyper-reality, for Lash (2002), who draws on him, is a more positive order of 'information and flow'. For Baudrillard the question of social bonds and *contract* have been replaced in the 'order of simulacra' by mediating forms that permit merely *contact*. He suggests that

> the rational sociality of contract, dialectical sociality (that of the State and of civil society, of public and private, of the social and the individual) gives way to the sociality of contact...thousands of tactical combinations. But is this still a question of the *socius*? (Baudrillard, 1983: 83).

Contact as opposed to any residual sense of co-presence imagined by *contract* denotes a technologically permeated world that renders all human life as strategic, impermanent *combinations*. Combination suggests that twoness is not qualitatively different from oneness: the principle heresy within sociology! Baudrillard answers his own question about the fate of the socius some years later,

> With the Virtual we enter not only upon the era of the liquidation of the Real and the Referential but that of the extermination of the Other. It is the equivalent of an ethnic cleansing [...] If information is the site of the perfect crime against reality, communication is the site of the perfect crime against otherness, (Baudrillard, 1996: 109).

The kind of media ordered, and technologically permeated, reality Baudrillard imagines here absorbs otherness in its relentless production of *mere* identity and difference. The body is a social residuum which appears in the earlier work as something that merely reacts. Lash's writing is altogether more positive.

Lash (in Gane, 2004) suggests that if *The End of Organised Capitalism* (Lash and Urry, 1987) had been about fragmentation then *Economies of Signs and Space* (Lash and Urry, 1994) had been about de-differentiation and the need within sociology to understand the contemporary world in the form of mobile objects, mobile subjects and flows. Drawing intriguingly on Simmel and Baudrillard, they describe the 'fleeting, intense and diverse' character of social interactions in the Metropolis of modernity. Modernity provides for the fleeting, highly mobile and speeded up circulation of 'capital, labour commodities, information and images' (Lash and Urry, 1994: 12). This circulation can be thought of in terms of 'flows'. The viewpoint is clearly developed from the precedents in urban geography and history on the one hand and on the other the geographical-inspired features of Giddens' (1979, 1981) structuration theory, namely time-space distanciation. The latter is key to Giddens' project in establishing the features of social interaction within modernity, their relationship to the technologies that colonise them, and the experiential issue of the 'emptying-out of the present through the 'disembedding' of co-present aspects of the social which we have seen above occupied Giddens. In picking up the Baudrillardian theme Lash and Urry point out that precisely what modernisation consists in is the development of the *functional utility* in objects at the *expense* of the symbolic 'co-presence'. They express this principle, for example, in the in the following passage,

> What is increasingly being produced are not material objects, but signs. These signs are primarily of two types. Either they have a primary cognitive content and thus are post-industrial or informational goods. Or they primarily have an aesthetic...content and they are primarily postmodern goods [...] This is occurring not just in the proliferation of non-material objects which comprise a substantial aesthetic component

(such as pop music, cinema, magazines, video etc.) but also in the increasing component of sign value or image in *material* objects. This aestheticisation of material objects can take place either in the production or in the circulation and consumption of such goods. (Lash and Urry, 1994:15)

Postmodernization for Lash (2002) means the replacement of social structures by information and communication structures. These latter structures are whatever 'frames' 'flows of information, communications, images, money, ideas and technology' (Lash, 2002: 28).

Flows are quite 'real' in their effects producing for Lash 'live and dead zones', places of more and less intense social relations and cultural activities, even an underclass defined as not having access to flows.

The 'information city', characterised by its flows, presents the embodied subject according to Lash (2002: 205) with a *problem of navigation*. We should probe this idea further as the forms of stability and wandering that it portends are key to this kind of sociology. The order of the problem of navigation posed in contemporary high modernity is not the same as that presented to earlier moderns such as that form of self coherence strategy enshrined in Walter Benjamin's flaneur. The flaneur reconstitutes coherence through allowing the Metropolis to become a contingent and surprising place in the activity of walking. Lash imagines the embodied subject as quite distinct from the ICTs with which they are become involved. He seems certain that the concept of interface is needed to characterise a distinction that should be maintained between them. In considering 'technological forms of life' he says,

I operate as a man-machine (sic) interface – that is, as a technological form of natural life – because I must necessarily navigate through technological forms of social life. As technological nature, I must navigate through technological culture. And technological culture is constitutively culture *at a distance*. Forms of life become forms of life at-a-distance. Because my forms of social life are so normally and chronically at-a-distance, I cannot navigate these distances, I cannot achieve sociality apart from my machine interface. I cannot achieve sociality in the absence of technological systems[.] (Lash, 2002: 15)

In order to understand, then, in what ways the traditional grounding conditions of co-presence become absorbed into 'flows' that turn the problem of self *coherence* into a problem of *navigation* we need to have an idea of what the medium of navigation is. However, in the picture so far presented there is no more said than what can be covered by the rather traditional idea of subject-body-skill-technology. This medium of flow appears to consist of two connected ideas that give liquidity to the high modern condition: networks and signs. For networks Lash provides us with an image that is in some tension with the idea of the embodied subject before an interface,

> Networks are the sites through which the flows (of money, images, utterances, people, objects, communications, technology) navigate...For [Deleuze and Guattari] most important are flows, 'pulsions' of desire and 'lines of flight'. These flows gain hegemony in the general 'de-territorialisation' of structures and institutions. But there is never the pure indifference of flows. The de-territorialised flows wind up 'solidifying' in a group of new 're-territorialisations', some of which become infrastructures for the flows themselves. Networks and actor-networks are such re-territorialisations. (Lash, 2002: 205)

In this view the problem of navigation, what the problem of coherence has now become, depends on what 'making links' means. The sociology of flows renders the activity of making links problematic to human life. Or put another way, the condition of flows entails that the human act of 'making' shows up as 'making *links*'. Making links in itself, and the problems it poses, is not a new idea. The traditional view of the skilled actor is imagined here as embodying capacities that enable engagement with an 'object-world'. Indeed, this way of viewing the subject-body-object field invites the Heideggerian connections that have appeared in sociology many times before, and have reappeared again more recently with renewed vigour. However, there is now a new twist which the concept of flow suggests. Making links is tantamount to re-territorialising. But the basis on which this is achieved is one where the linkage itself seems to become the primary object. The traditional view of skill as embodied capacity relies on a figure-ground view of skilled performance where the activity of making the links is invisible relative to what is linked. While this still has to continue to be the case, the implication of flows is that the background is routinely fore-grounded. We become sensitive to the flow itself and of ourselves in the context of our own complicity with it. In a curious parallel perhaps with Marshall McLuhan's famous phrase, 'the medium is the message', what coherence has become in the sociology of flows is a situation where my grasp of the *flow itself* is somehow more important than *what flows*. Indeed, ICT networking devices, such as the Internet mobile phone, gain their value not from 'content' but from their re-territorialising of concerns. Locating the whereabouts of one's children, tracking today's urgent transfer gossip for one's football team, taking a photo, storing a memory, retrieving data, receiving weather and transport reports, planning and re-planning one's journey while on the move, following an ecological or human disaster to which one has formed a commitment, sending work e-mails become 'de-distanced events'. My sensitisation to my own concerns are made contingent in an embodied way on the way the phone permits a condensation of the managerial problems posed by flows. This is not to say that locating one's children is less important than managing the information of their whereabouts! Rather, that the contingency and responsivity of those concerns are complicit with, produced by and show up with the manner in which the flow has now become managed, or navigated, and condensed with other concerns. This seems to be the meaning of 'navigation'. Condensing flows in this manner is not the same as co-presence. By attending to the condensation of flows themselves I construct strategies of coherence that promote sociality always at a distance.

It is in Lash's discussion of the sign that draws him closer to Baudrillard. Indeed, it seems that it is in the context of the sign that 'flow' gains its currency and dynamism. In order to grasp the new stabilities and mobilities constituted in this field, we need to understand the connection between embodiment and technology as *commutative*. The concept of the sign is something which already circulates, is already mobile and already flows. Baudrillard and Lash's conception of the sign emerge from a critique of both Saussure and Marx (which there is no space here to discuss in detail). It is important to the problem of stability and wandering which has become transmogrified in the context of the discussion of flows. Signs, like money in the traditional Marxian analysis, become the medium of social transformation by virtue of their capacity as a medium of exchange within a technosphere increasingly designed to handle them and promote their 'liquidity'. Signs are also central to our own experience of ourselves and the means by which we constitute self-understanding, stability and coherence. Lash and Urry (1994) discuss this in the context of extending the 'emptying out of meaning', first witnessed in the context of the objects of modernity under the conditions described by Marx. However, through the nature of the sign the same fate is in store for the subject of modernity by extension of the logic of circulation and commutation. This seems to be the sense in which, via the absorption of the social, the new form of loss of self is to be constituted. Rather than the 'squandering' self in prodigal mode, the potential is for a self 'squandered' in the service of flows. The mechanism for this would appear to be the manner in which embodied concerns are made contingent, by practices of linking, on a sensorial link with the flow itself. It is this link that technology foregrounds.   Once again, the embodied subject is reconfigured as a kind of 'remainder'. The 'territory' into which one wanders is a flow, but so is the territory from which one came – being a remainder becomes endemic. One's coherence is a striving to condense flows, not a journey between them.

Conclusion

I have tried to produce an account that sets the perennial human problem of stability and wandering in the context of the literatures that worry about them. Where that literature is a sociology attempting to come to terms with the 'fading of the social' and the increase of mobility and flux, then stability and wandering take place at a new threshold – a 'threshold of indistinction'.   The threshold needs to be considered as a field of embodied activity in the 'technosphere'.   I have suggested that Habermas's account of the 'fading social' has at its core an unexplored region of sensorial and embodied activity where his key issue of the loss of the social and human coherence needs to be taken back. This runs counter to readings of Habermas that say his approach lacks a sense of embodiment: it is based on it, though not developed. I argued that the core concepts drawn upon by Lash arguing the 'absorption of the social' depend similarly on understanding transformations taking place at a threshold of indistinction. These transformations to what now constitutes our stability and wandering take place within our very acts of making. Again a sensorial domain has been left undeveloped. The self, for its coherence, is reliant on forms of embodied responsiveness within the technosphere. Despite the fact that, now, unlike Odysseus, we will never reach 'home' and indeed stability may have become a function of wandering. But it is comforting to assume that these age-old themes remain.

Acknowledgements

I am grateful to Wendy Bottero, Graham Crow, Paul Dowling and Susan Halford for helpful discussion on the arguments in this chapter. Any incoherence in it is my own.

References

Agamben, G (1998) *Homo Sacer: Sovereign power and bare life*, California, Stanford University Press

Baudrillard, J (1983) *In the Shadow of the Silent Majorities*, New York, Semiotext(e)

——. (1996) *The Perfect Crime*, London, Verso

Bauman, Z (1993) *Postmodern Ethics*, Oxford, Blackwell

——. (2000) *Liquid Modernity*, Cambridge, Polity

Beck, U (2000) *What is Globalisation?*, Cambridge, Polity

Byock, J (1990) *The Saga of the Volsungs*, Harmondsworth, Penguin

Castells, M (1996) *The Rise of the Network Society. The Information Age: Economy, Society and Culture* Vol. 1, Oxford, Blackwell

Cregan, K (2006) *The Sociology of the Body : mapping the abstraction of embodiment*, London, Sage

Crossley, N (2001a) *The Social Body: Habit, Identity and Desire*, London, Sage

——. (2001b) Embodiment and social structure: a response to Howson and Inglis, *The Sociological Review*, 49 (3), 318–326

Dowling (2001) Social Activity Theory (working paper) http://homepage.mac.com/paulcdowling/ioe/publications/sat/index.htm

Gane, N (2004) *The Future of Social Theory*, London, Continuum

Giddens, A (1979) *Contemporary Problems in Social Theory*, London, Macmillan

——. (1981) *A Contemporary Critique of Historical Materialism*, London, Macmillan

——. (1990) *The Consequences of Modernity*, Cambridge, Polity

Habermas (1987), *The Theory of Communicative Action* Vol. 2, Cambridge, Polity

Howson, A (2004) The Body in Society: an introduction, Cambridge, Polity

Howson, A and Inglis, D (2001) The body in sociology: tensions inside and outside sociological thought, *The Sociological Review* 49(3), pp297-317, 2002a

Lash, S (1999) *Another Modernity, A Different Rationality*, Oxford, Blackwell

——. (2002) *Critique of Information*, London, Sage

Lash and Urry, 1987 *The End of Organised Capitalism*, Cambridge, Polity

Lash, S and Urry, J (1994) *Economies of Signs and Space*, London, Sage

Lyng, S (2004) Crime, edgework and corporeal satisfaction, *Theoretical Criminology* Vol. 8(3): 359-375

Held, D., McGrew, A., Goldblatt, D. & Perraton, J. (2002) *Global Transformations: Politics, Economics and Culture,* Cambridge, Polity

Maffesoli, M (1996) *The Times of the Tribes: The decline of individualism in mass society,* London, Sage

Mead (1934), *Mind, Self and Society,* Chicago, University of Chicago Press

Meeks, W (1999) *The Moral World of the First Christians,* Westminster, John Knox Press

Tambling, J (1990) *Confession: Sexuality, Sin, the Subject* , Manchester, Manchester University Press

Thompson, J (1995) *The Media and Modernity: a social theory of the media,* California, Stanford University Press

Urry, J (2002) *Global Complexity,* Cambridge, Polity

Vass, J (1996) Economic Socialization: a tourist in my own transactions' in *Research into Children's Economic and Social Understanding,* Vol.5

—. (2002) Social Strategies in the Discourse of Societal and Citizenship Understanding to appear in Ross, A (ed) *Future Citizens in Europe,* London: CiCE

CARNIVALIZATION, THE BODY AND THE LIMINAL: FROM THE FLANEUR TO THE URBAN PRIMITIVE

LAUREN LANGMAN

Abstract: The religious practices of antiquity typically involved various rituals of frenzy, abandon and Dionysian passion valorizing the body and extolling ecstasy, drunkenness and sexuality. Consider the worship of Ishtal, Baal and even Venus. Judaism, as an ascetic patriarchal religion, demanded the repression of desire and disdain of the profane and dangerous body. The Dionysian was marginalized, relegated to the liminal. In Christianity, the repression of the profane body and bodily became sacralized and intertwined with doctrines of sin and salvation. Repression has always been dialectical; it was tolerated in the peasant carnivals of the Middle Ages that preceded Easter. With the ascent of modernity, as a rational, capitalist, bureaucratic order, the repression of the bodily endured even when the power of religion waned. With the globalization of capital, we have seen a number of adversities ranging from economic distress to questions of meaning. But we have also seen a 'transvaluation' of the ethical and the reemergence of various aspects of the carnivalesque as intrinsic moments of the 'culture industries' in which the various expressions of transgression have become intertwined with commodification. This carnivalization can be understood as 'repressive desublimations' that both attempt to recapture long dormant religious traditions, as well as deflect or contain the discontents of global capital.

Introduction

The Dialectics of Urban Life

The city, as a location with its unique discursive practices is both repressive and emancipatory, social density is constraining, yet the anonymity of the city frees the person to pursue a multitude of options. The site of this dialectic is the body as both inscribed by various discourses, and at the same time, impelled by desire, the body is the basis of agency. This can and will be illuminated by the contemporary 'urban primitives' who exist within the city, yet embrace a primitive imaginary articulated in the embrace of various forms of tattoos, piercings, scarifications etc.

The rise of civilization began with the emergence of the city and indeed both share the same linguistic roots. As early astronomers charted the seasons and engineers fashioned irrigation systems, cities began to grow as centers of trade, governance and/or religious/cultural centers. But with ever larger and growing populations, often consisting of peoples from near and far, traders and merchants, scholars and travelers, strangers close and distant, the city has always been the center of social and intellectual diversity and innovation as different ideas and cultures mixed. With the very emergence of the city, full time literati not only administered empires, but religious virtuosos, transformed the folk cultures of religious myths, tales and practices into formal theologies with cosmologies, eschatologies, ethical codes. That the theologies and religious practices articulated by intellectual elites have disciplined the body to sustain the domination of economic elites has an essential premise of the Enlightenment. Further, following Hegel, and of course Marx, systems of domination foster resistance and indeed contestation. Thus for much of written history, cities have been centers for the administration and legitimation of domination. Yet at the same time, they have also been the sites where various forms of resistance, contestation and challenges have been born.

Modern urban life, as the seat of the rational, 'money economy', has long depended on repression. Soldiers, militias and men-at-arms have attempted to police the city, and punish and/or isolate deviance. Internalized repression has been needed to maintain the methodological orientations of everyday life oriented to goal-orientated work that depends on a particular the character structure disposed to engage in that kind of work, eg the Protestant ethic of self control and denial, a 'this worldly asceticism'. Otherwise said, as Nietzsche noted and Freud affirmed, 'modern civilization' rested on the suppression of the passionate Dionysian in order to sustain the Apollonian, the controlled, orderly harmonious. For Nietzsche, the Christians of the early Roman Empire, marginalized, urban artisan classes, were filled with *ressentiment* toward the rich, powerful and licentious Roman. They found compensation in a 'transvaluation of the ethical' -what had been 'bad' became good, that is the values of strength, power and bravery had become inverted and submission, charity and humility were exalted to provide the subalterns of Rome with 'moral superiority' and guaranteed entrance to heaven. Yet by the 4th C., the Romans themselves had embraced Christianity. But as

Nietzsche argued, long after the demise of Rome, and even after Christianity was challenged, following the Enlightenment based modernity, the 'slave mentality' endured beneath the 'tyranny of Reason', fostering conformity, petty banality and the repression of desire that led to the sickness of self-hate and contempt based on feelings of inferiority, weakness, and jealousy.

As Stauth and Turner (1986) argued, Weber was much influenced by Nietzsche and saw modern rationality as a thin veneer of goal oriented rational action, *zweckrational* in a tenuous relationship to bodily based, affective action and tradition. Protestant gave rational action a theological cast by sacralizing the control of bodily desire, sexuality and aggression, through sublimation into work and a methodological stance to everyday life oriented to long term goals. Freud took this as a trope for the very rise of civilization in which primitive bands felt guilt and remorse after the patricidal act that began the ascent of repression. Freud (1930) suggested that civilization was based on repression embodied by an anal compulsive character structure in which sphincter control becomes the model in which bodily desires were first repressed and then sublimated into attachments to other people and work for the common good, but at the cost of personal happiness and fulfillment.

But Freud's analysis was quite a-historical. Nor did he explain the sociological processes of just how and why people would renounce the pleasures of instinct gratification in favor of the demands for renunciation and long-term symbolic gratifications. Enter Elias (1978) who located socially required modernity in a historical process in which the aristocratic courts of Europe first began to embrace manners and politeness as integral moments of courtly society and diplomacy. The courtly embrace of manners would eventually establish the underlying repression overt social styles of the bourgeoisie.

The Modern City

The classical models of the city first sketched by Simmel, Worth, Park and Burgess emerged in an age of steam powered factories and railroads, department stores and masses of urban proletariat. European cities such as Berlin, Vienna, London and Paris, and US cities such as New York, Chicago and Boston were the centers of the new industrial era, the sites of its markets and finances and its cultures of refined civility. But at the same time as the growing cities began to build libraries, museums, opera halls, symphony halls and the other accouterments of civilized culture-read the

'distinction' of the 'haute bourgeoisie', certain social processes began that
would contest the dominant ethos.

1.   The art and music of the bourgeois began to challenge the received
        repression. Perhaps this might be seen when Manet's Olympia
        portrayed the nude body of a prostitute empowered by her gaze at
        the viewer and the transgression of her subjectivity. She was
        without any shame over her nudity or her line of work. Gaugan's
        erotically free maidens of Polynesia valorized the untainted
        'primitive' as a 'noble savage' whose freedom and eroticism were
        untainted by bourgeois repression and hypocrisy.
2.   With growing urban populations and improved forms of
        transportation, widened boulevards and Metros, after Hausmann we
        began to see the rise of department stores, gallerias and arcades.
        While nascent, we began to see the origins of consumerism and life
        styles based on an elegance of consumption, eg the *flaneur* (urban
        dandy) who turned his, and it was mostly his, life into a work of art
        seeking grandiloquence in style, grace and sophistication. He,
        according to Simmel, was a pure form of urbane civility, much as
        he himself was an ideal typical *flaneur*, and the only *flaneur*-
        sociologist to inspire the field for over a century.
3.   Little noted at the *fin de siecle* of the 19th C. but new technologies
        of communication and entertainment were emerging, eg the
        telegraph, phonograph, radio and film. By the 1920's a mass
        mediated popular culture was emergent that would become highly
        intertwined with marketing, advertising and consumerism. While
        the growth of consumption based life styles was slowed by the
        Depression and then WWII, the following decades would witness
        the proliferation of consumerism extolled by the growing media of
        television. Advertising and marketing began to decry the illnesses
        of 'civilization' and promised that with the purchase of certain
        products, the person would find health, beauty, fame, fortune, love
        and a better life-free of bad breath, body odor or headaches.
4.   As these processes were taking place, consumerism required the
        relaxation of the controls, constraints and repression that enabled
        industrialization to emerge, but such as culture of constraint would
        be unlikely to promote the impulsivity needed for a consumer
        society. Thus advertising, often disguised as movie plots, began to
        craft valorized identities based on consumption. Otherwise said,
        everyone with money, or at least credit, could be an urban *flaneur*-
        or at least its mass produced versions. But what needs to be noted is
        that consumer based identities that yield privatized hedonism, by

withdrawal of cathexes from civil society and political institutions, serve to reproduce the adverse conditions that consumerism would ameliorate.

5.   But as these events took place, so too in the various liminal sites and interstices of urban life, a culture of inversion and transgression emerged catering to more 'popular', indeed 'vulgar' tastes. The In face of art museums, came freak shows and museums of oddities. For many of urban masses, amusements parks such as Coney Island or Riverview provided escape and respite from urban life. (It also gave young people spaces free of surveillance where young couples might have a few moments to kiss, grab and grope without panoptic parents or priests.

These interconnected moments would create liminal times and places at the margins of the city have long been places where the anonymity of the city provided freedom from social control where the otherwise prohibited and proscribed might be enjoyed. Thus the city as the site of rationality and repression also led to liminal realms of indulgence and inversions ranging from illicit sexuality to homosexuality.   Thus besides the workplace emerged the 'funplace', a world of bars, bordellos and the bizarre. Otherwise said, along with the production of goods and services, so too did the modern city become major site of consumerism by providing entertainment zones, offering various games and pleasures-including the otherwise prohibited that were valorized in the liminal sites. The 'fantasy city' is an enchanted dreamworld of tourism, entertainment and consumption where every act is permitted and every thing is for sale (Ritzer, 2000; Hannigan, 1998).

Following V. Turner (1966), a given social structure as an organization of constraint, dialectically fosters alternative realms apart from the typicality of the quotidian. There is a bifurcation of reality into the normative structure and a liminal anti-structure. The liminal stands apart from the usual as illness stands in opposition to the typicality of health. Liminal realms are times and sites of freedom, agency (empowerment), equality, license and spontaneity. The liminal realm is often the site for resistance struggles, inversions and repudiations, indeed flouting norms and expressions of acts or feelings that are usually forbidden or taboo. The liminal culture of the ludic grants pleasurable moments of indulgence and resistance- if only for fleeting moments, if only in marginal, interstitial or even imaginary sites of otherwise prohibited gratifications of inversion and license, reversals of norms and. There is a suspension if not reversal of usual codes of morality, dress and self-presentation. All that was otherwise prohibited was celebrated. The liminal is the place for the Dionysian that was suppressed by the Apollonian. For a limited time in a liminal site, there could be 'transvaluations of ethics' in which alternative meanings could be negotiated.

The Global City

The modern city has been radically transformed in the last few decades. In the late 60s, as urbanologists began to write the epitaphs for a dying social space, as most people began to head to the suburbs, a funny transformation happened on the way to the shopping mall. More specifically, little noted at the time, but as a result of computerization, miniaturizations and digitalization, along with the expanded uses of the APRANET (now the Internet), a number of factors led to the globalization of the economy. Globalization, the defining moment of the present age, has transformed the world as we know it. Techno-capital has enabled the emergence and proliferation of a de-territorialized economic system, decoupled from any particular nation state. Capitalism, with advanced technologies of information, production, finance and distribution, has now become a re-spatialized system dominated by Trans National Capital.

But as Sassen (1998) and others have noted, the re-spatialization of global capital has created 'global' cities that exist as nodal points in the new global economy. This new economy, based in large part on rapidly changing advanced technologies of information and communication to control, command and coordinate networks of research, production, distribution and finance. Every day vast sums of money and information flow across the world, typically from one global city to the next. Indeed these cities often have more commerce between them than they do with their own hinterlands. Thus the city still requires large concentrations of workers, but one segment of this work force, the elite fractions of transnational firms consists of various advanced specialists and professionals in banking, accounting, law, etc. Further, these workers often require large numbers of support from experts in engineering, computing, international law etc. But most important, these workers tend to be quite affluent and therefore require the services of a large number of other workers, little skilled, little trained and little paid. These are the workers who do the janitorial work in the skyscrapers, prepare the food for the meals these people eat, they take care of their lawns, clean their homes and take care of their children.

A final point needs to be made, the most rapidly growing segment of the new economy is tourism and travel. Accordingly, urban economies, catering in part to the affluent workers in the global economy, have become more and more based on entertainment and tourism, and indeed, global cities compete with each other by offering various forms of entertainment from art shows to theatrical presentations to music ranging from rock concerts to blues bars. Thus global cities as 'entertainment zones', providing various 'urban amenities' tend to not only tolerate the liminal, but these realms of commodified transgression become integral to urban economies (Florida, 2002; Clark, 2004).

Thus an increasing part of the tourist trade is devoted to the transgressive. Thus for example, the Mardi Gras of New Orleans features various street displays of breasts, genitals and even an occasional public sex act. Carnival in Rio, has become a primary tourist destination with its promises of sex and fun. Cancun offers sex, beer, bare boobs and more sex for college students. Of course, cities like Bangkok and Manila have become destinations for sex-tourism. Cities like Sydney, Johannesburg and San Francisco are major centers of gay subcultures. Otherwise said, with globalization, cities have become entertainment zones providing all kinds of amusement based consumption from the clean and wholesome shopping centers to the liminal, transgressive and inverted. Among the growing subcultures of inversion, we have seen the growth of an anti-modern, anti-urban sub culture of 'urban primitives' or 'modern primitives' who reject the values and life styles of modernity and would return to the imagined past of pre-history.

Globalization, Surplus People and Subcultures of Resistance

Globalization, with its neo-liberal ideology universalized, new forms of post Fordist, flexible production and management (TQM), and control of national governments, has created unprecedented wealth. A new class of trans national elites has emerged that has garnered the lion's share of this wealth (Korten, 2002; Sklair, 2001). But the distributions of that wealth have been sharply skewed and most of the newly created wealth has gone to the elites. (Perruci and Wysong, 1999). There has been stagnation for the many and downward mobility for some. Otherwise said, one of the main consequences of the re-structuring of the political economy has been the production of surplus populations available for the growing numbers of lower echelon McJobs in the service sectors of the global cities. The growing inequality, together with more 'flexible' management style of organization, eg TQM, has meant that for most people, even those with college educations, more and more are likely have lower echelon service jobs, of limited duration in which work has become a series of short term projects that provide little intrinsic gratification (Sennett, 1998).

Many youth, facing bleak prospects in the work force, have opted out altogether. For some, there are the typical alternatives of crime and/or addiction. But others seek membership in compensatory alternative subcultures of entertainment as resistance such as heavy metal, hip hop, grunge, punk, or goth, etc that reflect the many tastes of youth culture. Of course we mean the lure of Marilyn Manson, Eminem or Ozzie Osmond. Adolescent autonomy, rebellion and disdain of parental values and life styles and quests for an independent identity has been a hall mark of youth cultures since the early 20th C. Most youth cultures, from *avant-garde* bohemians, to existentialist beats, folkies, rockers, hippies etc. provide a 'psycho-social'moratorium in which a variety of identities and life styles of resistance and rejection could be explored, embraced and modified as preparations for adulthood. At the same time, we should note that the more 'deviant' the alternative identity, the more likely it becomes that only in 'global' cities or their environs where there are large concentrations of people and spaces of anonymity.

Identities of resistance, as disdain of the dominant culture, publicly repudiate the norms, values and life styles of the larger society, often seek the set the person apart in terms of cultural tastes expressed in fashion, adornment and self presentation. Various styles of dress, coiffure and ornamentation valorize the vulgar, the obscene, the shocking and the grotesque set the person apart from most others -and incorporate him/her into a subculture of resistance. Fashion, and adornment, as an immediate advertisement so self and statements of identity, has become an important marker that can make a statement of opposition, rebellion and resistance-confrontational dressing aimed to offend (Wilson,1987).

What is especially interesting about contemporary forms of fashion and adornment has been the decoration of the body that has itself become a template upon which aesthetic sensibilities are inscribed and through which selfhood is articulated. More specifically, in the past few decades, as globalization has reduced the opportunity structure for many youth, we have seen the rise of alternative subcultures. Body modification, tattoos, piercings, surgical modifications (split tongues, implanted horns) are fashion statements indicating a growing moment of resistance, rebellion and distinctive marker of inclusion. Thus perhaps 1/3 of youth are likely to have some form of body piercing besides earlobes such as a stud, post, or ring, and quite often a tattoo.

The tattoo has had an interesting history. It was first embraced by sailors in the 19th C who were trading or whaling in various parts of the South Pacific from Polynesia to Australia and New Zeeland who encountered various 'primitive people'. It was embraced by urban working class males as symbols of masculinity, devotion to a woman, and/or individuality (Cf. DeMello, 2000). It would also become embraced by soldiers and sailors as a badge of patriotism. It would also become a marker of subcultural identities such as bikers, gang members and/or prisoners. While each subculture might embrace a different aesthetic, that would change as large numbers of more affluent, educated youth would embrace body modification as an aesthetic.

While at one point body modification was been considered quite 'deviant', it is now so common as to be 'ordinary'. Nevertheless, there are extremes of body modification that push the envelope from the fashionable to the curious to the grotesque. We are of course referring to multiple rings, studs, posts etc, various implants such a horns or furrows, split tongues, wide spread body tattoos, and perhaps most extreme, various genital piercings, implants, urinary reroutes, modifications and saltings (injections of saline solutions into the scrotum or labia). While many aspects of youth culture are eventually shed once youth enter the 'adult' worlds of work and responsibility, many of the body alterations are such that these youth become quite unlikely to re-enter the main streams of the dominant culture. Few corporations, professions or public service companies are likely to hire a person with protruding horns, a split tongue, septum ring and a variety of facial tattoos. Nor might such folks consider such 'respectable' life styles.

The concerns with body modification speaks to a number of sociological questions. Modern sociology has rediscovered the body, especially following the work of Foucault (1977), B. Turner (1996) or Shilling (1993). Shilling argues that the body can be seen as socially constructed. For Foucault (1977), body has been seen as the site where discursive power, mediated through disciplinary practices inscribe certain forms of subjectivity. The gaze, the look, enables power to define and control. But on the other hand, the 'naturalistic' approach suggests that the body provides the basis for agency and self articulation. We might also note Elias (1978) historically charted the body as the site of the 'civilizing process', understood as progressive regulation of bodily desire and bodily shame became systematically repressed and controlled by ever more repressive standards of modesty and manners. But this repression enabled rational capitalism, Nation States and 'civilization'.

These various expressions of body modification can be understood as essential moments of the resurgence of the medieval carnival, whose essential features celebrated transgression, the vulgar and grotesque. Its contemporary forms would recapture a lost moment of an imagined 'premodern' primitive Dionysian moment free of the Apollonian repression. As Bakhtin (1968) noted, the carnival as an inversion of what is 'normal', valorized the ordinarily prohibited. It celebrated the lower body whose orifices and products were foul, vile and repugnant. Many adherents of body modification, esp tattoos, burnings, scarrings, etc, regard their embrace of the grotesque as a rejection of the sterility and emptiness of modernity and label themselves as 'modern' or urban 'primitives' (Vale and Juno (1989). Otherwise said, we would like to suggest that body modification be understood as a transitional marker of passage from the dominant culture though a 'de-civilizing process' that would restore a lost moment before Judeo-Christian religion displaced various pagan, fertility religions that valorized agenic, sexual body where desire unbound. This is expressed as a transvalued grotesque of the body as a statement of active disdain, resistance and rejection of the repressive sterility of modernity and instead authenticity regained.

Part I: Political Economy and Identity

From Carnival to Capital

For Weber, the emergence of modernity, culturally understood as the 'disenchantment' of the world', first emerged along with rational commercial practices and more rational forms of governance. The rise of trade led to the rise of the city and as the city disposed freedom and rationality, commerce became more rationalized. Ascetic Protestantism emerged that would demand the repression of desire to dispose an 'inner determination' to make one's work a 'calling', a sacralized *beruf*, understood as a long term career in which there was a methodical orientation to everyday life. Thus with the rise of the commercial city, the routinized economy, routinized organization emerged along with a character disposed to routinization. One way we can we can trace the rise of modern, rationalized capital can be seen in the rise and fall of the carnival.

The carnival of feudal Europe first emerged as a moment of peasant popular culture that stood apart from, if not opposed to the official feasts and tournaments that celebrated and secured the power of the aristocratic elites (Bakhtin, 1968). The carnival centered on festive rituals and practices, located in alternative times and sites in which typical patterns of hierarchy, deference and demeanor were ignored (Bakhtin, 1968). These were times and places of total indulgence in wine, song, dance and sex. The typical restraints of everyday life waned, carnivals were times for the systematic transgressions of boundaries (Bakhtin, 1968).

The carnival was a site of resistance apart from everyday life and subservience to the landowners and clergy. All that was proscribed was celebrated, boundaries of 'decency' were now valorized and celebrated, especially the bodily, bodily indulgence, the orifices, excreta, the profane, the passions, the vulgar, the grotesque and obscene. What might be otherwise disgusting and grotesque was valorized. A central theme for the carnivalesque was the grotesque body. According to Bakhtin (1968), the genre of the grotesque stood in direct opposition to all medieval forms of high art and literature. Through mockery and satire it challenged the authority of the Church and state. For Bakhtin (1968), one of the most important aspects of grotesque realism is its function of degradation. He explains, degradation here means coming down to earth, the contact with earth as an element that swallows up and gives birth at the same time. To degrade is to bury, to sow, and to kill simultaneously, in order to bring forth something more and better. To degrade also means to concern oneself with the lower stratum of the body, the life of the belly and the reproductive organs; it therefore relates to acts of defecation and copulation, conception, pregnancy, and birth. Degradation digs a bodily grave for a new birth; it has only a destructive, negative aspect, but also a regenerating one. To degrade an object doe not imply merely hurling it into the void of nonexistence, into absolute destruction, but to hurl it down to the reproductive lower stratum, the zone in which conception and a new birth take place. Grotesque realism knows no other lower level; it is the fruitful earth and the womb. It is always conceiving (1968, p. 21).

In the carnival, the people's laughter is the materialization of the degradation of authority. Thus laughter symbolizes the collective comprehension and shared affirmation of the satire-it marks resistance to the official life and celebration of the utopian moment of the carnival.

The grotesque body transcends physical boundaries and inverts appropriated standards of gender and hygiene. It is embodied in the characters of Gargamelle and Gargantua. These gluttonous, gruesome bodies are brought into being by the processes of grotesque, sexual consumption and excretion (e.g., defection, masturbation, etc.), which were not typical characteristic bodily functions of dignified nobility. The scene of Gargamelle giving birth to Gargantua, which ends in her death, epitomizes the grotesque image of the body. The principle concepts of this scene, the body's vulgar overindulgence and consequential obesity are akin to the thematic concerns of grotesque realism. This miraculous event takes place at the festive banquet, which resembles the days of feasting at Mardi Gras, where Gargamelle, after gorging herself with tripe, the intestines of fattened oxen, begins her labor immediately after dropping her right intestine. Her fleshy, overblown body cannot support the sickening, gigantic portion of food that she devoured and her intestine falls out. Bakhtin explains its significance: Bowels, intestines, with their wealth of meaning and connotation are the leading images of the entire episode. In our excerpt these images are introduced as food: *gaudebillaux,* an equivalent of *grasses tripes,* the ox's fatty intestines. But Gargamelle's labor and the falling out of the right intestine link the devoured tripe with those who devour them. The limits between animal flesh and the consuming human flesh are dimmed, very nearly erased. The bodies are interwoven and begin to be fused in one grotesque image of a devoured and devouring world. One dense bodily atmosphere is created, the atmosphere of the great belly. The essential events of our episode take place within its walls: eating, the falling-out of intestines, childbirth (1968, p.221-222).

It is a comic scene. As Gargamelle groans and wallows on the ground, a number of midwives rush to her assistance. Again, the focus is upon the lower strata and the turning upside down of norms, which becomes fantastically ridiculous as midwives mistake the dropped intestine for a baby and Gargamelle gives birth to Garagantua through her ear. The labor is consummated with Gargamelle's death and Gargantua emerging from his mother's ear womb with his first cry, which was actually a robust bellow, for a celebratory drink.

At first glance, one might argue that this discussion focusing on the lower stratum is in error, for the grotesque style is also concerned with the body's upper strata, specifically with the very important features of the nose, mouth, head, and ears, which become grotesque when assuming animal form. German scholar, G. Schneegans in a well-documented history of the grotesque, *Geschichte der Grotesken Satyre,* 'The History of the Grotesque Satire' (1894), makes reference to the caricature of Napoleon and the exaggeration of his nose. According to Schneegans, this representation of a human body part in the form of a snout or beak becomes grotesque when it reaches extraordinary proportions. Nonetheless, as Bakhtin (1968) points out, Schneegans fails to recognize that the underlying psychoanalytic meaning of the fantastic proportions of the nose is that it represents the phallus, thus the lower stratum. In congruence with literature from the Renaissance and the Middle Ages, carnival celebrations included ritualistic dances and short humorous sketches that supported the popular belief that the potency and size of one's genitals could be inferred from the size of the nose (e.g., the famous carnival 'Dances of Noses' of Hans Sachs).

The grotesque style is concerned with that which gapes, bulges and protrudes from the body. The eyes are grotesque only when they protrude from the head; otherwise, they are generally considered to be non-comical and an expression of the individual. According to Bakhtin (1968):

> The grotesque body is... is a body in the act of becoming. It is never finished, never completed; it is continually built, created and builds and creates another body. Moreover, the body swallows the world and is itself swallowed by the world (let us recall the grotesque image in the episode of Gargantua's birth on the feast of the cattle-slaughtering). This is why the essential role belongs to those parts of the grotesque body in which it outgrows its own self, transgressing its own body, in which it conceives a new, second body: the bowels and the phallus. These two areas play the leading role in the grotesque image, and it is precisely for this reason that they are predominantly subject to positive exaggeration, to hyperbolization; they can even detach themselves from the body and lead an independent life, for they hide the rest of the body, as something secondary...Next to the bowels and genital organs is the mouth, through which enters the world to be swallowed up. And next is the anus. All of these convexities and orifices have a common characteristic; it is within them that the confines between bodies and between the body and world are overcome: there is an interchange and interorientation. This is why the main events in the life of the grotesque body, the acts of the bodily drama, take place in this sphere. Eating, drinking, defecation and other elimination (sweating, blowing of the nose, sneezing), as well as copulation, pregnancy, dismemberment, swallowing up by another body- all these acts are performed on the confines of the body and the outer world, or on the confines of the old and new body. In all these events the beginning and end of life are closely linked and interwoven ((1968 p. 317).

Thus, the upper and lower body function together in the processes of consuming, excreting, and defecating, in which the regenerative process is always downward. The 'downward swing, which brings together heaven and earth. But the accent is placed not on the upward, but on the descent.' The downward movement is characteristic of the merriment of all grotesque festive forms. Clearly, this is seen in the processes of consumption and excretion, but the downward motion is also symbolic in the physical beatings. The victim of the fight is thrown to the ground and crushed to pulp by his opponent. Such is the case in the bloody beatings by Friar John. Friar John flattened noses, knocked out eyes, crushed jaws, broke ribs, bashed in chests exposing heart and lungs. He brained some and smashed the legs and arms of others.

Stallybrass and White (1988) noted how the fairs, carnivals and festivals of England were the dominant feature of feudal life. And in these festive rites, vast quantities of ale were consumed, and of course, there was mass debauchery. Yet with the growth of commerce between villages and town and rise of trading cities, slowly but surely, a more ascetic, orderly society emerged. This commerce including trade with countries like Turkey, so too did Europeans begin the consumption of coffee-the beverage of sobriety become popular. And as bourgeois commerce began to ascend, the carnival began to wane and recede, save in a few places like Venice. As coffeehouses began to spread throughout Europe, alehouses began to close, and slowly but surely, a civilizing process began to spread throughout Europe, perhaps catalyzed by Protestantism. Capitalist modernity, at least in its ascetic Protestant forms, would become a world transfomative moment with an ideology of work, individualism and self control that was consistent with the demands of the emergent capitalist political economy, but further, there emerged a character structure consistent with the new forms of work and value. But how and why was subjectivity transformed? Little has been said about the rise of the agenic subject, constrained to asceticism by inner controls.

It would seem as if a number of factors might be noted, starting with the value of work in the monastic orders, the individualization of character after the institutionalization of confession, printing, mass literacy, schooling and the emergence of childhood as a period of protracted socialization. A character controlled from within by shame and guilt, eg., a historically emergent super-ego became a factor leading to the rise of modern, capitalist society. As the feudal economy waned and social ties became attenuated, there was an embrace of Protestantism that would assuage anxiety over the new 'freedom', eg attenuated social ties and the erosion of the certainty of salvation promised by the Church. To alleviate the 'salvation anxiety', work, served as a counterphobic mechanism in which compulsive activity was both a palliative to the person and the requirement of work in the new economy. Moreover, with the emergence of childhood as a moment in the life cycle, these values, often guided by the Bible insured the intentional socialization of Protestant types who would not only reshape the world, but their descendents would see that world as a marketplace. Yet as that world would become a marketplace, so too would its values of restraint and asceticism wane as globalization fostered a consumer based 'amusement society' in its 'fantasy cities'-the nodes of global capital.

Back to Carnival-As Commodity

As we earlier noted, globalization has led to a combination of growing class inequality, masked by the unending stream of ludic carnivals and consumer spectacles offered by an entertainment society that is 'amusing itself to death' (Postman, 1986). This has led a world much reminiscent of feudal society. But this ascent has been little known or charted in face of escapes to 'hypereality' and/or what has been politely called 'the closing of the American mind', or less politely called 'dumbing down'. We shall endeavor to understand the present age as *carnivalization of the world,* consisting of two related moments. 1) There has been a *'re-feudalization'* of the class system in which a small minority of elites posses great wealth while the masses grow more poor. 2) Just as the *carnival culture* of the middle ages served to sustain an aristocracy of land, the mass mediated simulation of modern carnivals of consumption sustains the new aristocracies of transnational capital.

Campbell (1987) suggested the erosion of the Protestant ethic actually began in the Romantic era when authentic self fulfillment-in realms apart from the horrors of urban life-was exalted. The Romantic sought to express his/her 'true self' and individuality apart from the realms of dehumanizing work in anonymous cities. Notwithstanding, this appeal could not find mass support, most people needed to work and could not afford the gracious lifestyles of 'country gentry'-nor their cultural legacies, the urban *flanneur*s. Yet as we noted, by the end of the 19th C, various currents of thought, especially Nietzsche, art and literature, (including a flourishing pornography), prefigured the rise of mass consumption. Yet that *mass* consumption would need *masses* of consumers with disposable incomes, and a *mass* media that would colonize consciousness to erode repressive asceticism and foster subjectivities articulated through consumer based life styles (Langman, 1992). The transformation of the industrial society of production to the post industrial society of consumption was a long, complex process that occurred at the material, social, cultural and subjective levels. This history has been told elsewhere but for our purposes would first note that urban life required families to purchase rather than produce what they needed, eg foodstuff, clothes etc. Thus prepared foods were increasingly 'branded' as Campbell's, Heinz or Kellogg's. With the growth of mass media, we began to see a three fold process in which 1) the capitalist system would need to expand it offerings and markets to sustain profitability, 2), and thus the a valorization of symbolic differences of the otherwise similar, and the arbitrary linking of products to desirable states and selfhood became the means by which advertisers would encourage consumption, 3) so that given the events following WWII, the universalization of television, and general economic growth, emergent consumer culture created an age specific market, the 'teenager', a newly created market for various forms of clothes, popular culture products and soon, the 'modern', sanitized entertainment sites such as Disneyland. Yet as shall be seen in part III below, for segments of the newly emergent youth cultures of subsequent generations, neither the available service jobs nor the superficialities of consumer culture could provide emotionally gratifying, identity granting communities of meaning.

Part II: Identity and Desire

Identity, as the reflexive scripts and narratives of a distinctive group or subculture is a mediating linkage between the individual, dynamically understood, and the larger society. Groups in turn provide the basis, contexts and socialization processes that provide the individual with his/her own individual identity as that provides a self-referential locus of subjectivity as the person moves through the flow of social interactions across time and space seeking affective gratifications (love, joy, pride) in the enactment of the routines of everyday life, and/or avoiding negative feelings, fear, anxiety, embarrassment, shame or sadness. Ones self is the experiential locus of gratifications and pleasure, as well as pains, frustrations, humiliations or insults. Otherwise said, people act in way that will hopefully secure emotional gratifications or avoid pain and frustrations. This is not a simple hedonism in that people will suffer great pains and deprivations to sustain and realize their identities. (We will see that various forms of body design and modification may be quite painful.) People: 1) seek membership in and attachments to identity granting 'communities of meaning' that provide social bonds and attachments; 2) they strive for recognition, self esteem and dignity granting statuses; 3) they exercise agency in striving to overcome powerlessness; and 4) given a universal fear of meaningless and death, people seek value system that provides explanations for pain, suffering and injustice, proposes, remedies and/or promises of amelioration, even if in the next life. (Langman, 2000).

1) *Community*: For most of human history, between the powerful bonds of an infant's attachments to caretakers, to family/clan work groups, to the periodic solidarity rituals of large groups, people are social creature with needs to be part of a community. (And communities also grant

identities, recognition and meaning-see below). For Freud, group life depended on 'aim-inhibited cathexes'. Later psychoanalysts like Bowlby and Winnocott, have seen early bonding as the foundation of healthy development. They claim people seek attachments-not instinctive gratifications. Similarly, sociologists have long argued that the collective rituals of family religion, national celebrations etc, or interpersonal rituals of manners and facework serve to sustain community, solidarity and engagement. At the same time, capitalism, under girded by Protestant individualism, served to attenuate social ties and connections. Today, between urban life, geographic mobility, the political-economic impacts of globalization, and the mass consumerism that individualizes people as consumers (even if as commodified pseudo-individualism) social ties are attenuated, if not rent asunder. This social fragmentation is often resisted. In this way, membership in communities of transgression can be seen as ways of establishing, kindling and affirming social bonds.

2) *Recognition:* Following Hegel's understanding of the master-slave struggle for recognition as the basis for self-consciousness, a number of psychoanalysts and social critics have noted the importance of recognition as the basis of self esteem; some consider the pursuit of self esteem, 'honorific status' much more powerful a motive than sexuality. And indeed far more men have died for the sake of honor in combat than for the love of a woman. Following observations by Sennett and Cobb (1972) echoed by Honneth (1995), and Taylor (1992), struggles for recognition/dignity, avoiding insignificance, may be seen as central to both the person and his/her groups.

3) *Agency:* A long line of philosophical critique, as articulated by Spinoza, Kant, Hegel and, Marx suggested that active self-constitution, striving for realization and creating reality is an inherent human tendency. With Nietzsche, philosophy moved from epistemology to psychology. *Will*, now framed as agency, became a central quality of investigation. For Marx, alienation the consequence of wage labor, externalized production. Commodity production stood as an external power, thwarting his/her agency, his/her humanity and selfhood. The humanly constructed systems of domination refluxcd back upon the person to render him/her powerless. To understand humans as willful is to see people as active agents, who strive to make their everyday lives meaningful and to attain certain long-term goals and values.

Powerlessness, the denial of agency is generally an extremely potent motive for individuals and groups. To be powerless is often humiliating and shame can be a powerful motive that fosters attempts to gain empowerment, to overcome domination. Indeed from slaves cutting of limbs to escape the chains of servitude, to masses enthralled with the empowering promises of a dictator, people seek empowerment. As Nietzsche suggested, to accept domination, to assent to slavery, leads to the sickness of a revenge denied. So too does fundamentalism dialectically foster subordination to power by establishing 'micro-spheres' of empowerment that provide illusory realms of power. But at the same time, the transgression of norms gives one a power over them

4) *Meaning:* People have a fundamental need to make their lives meaningful, they need frameworks of meaning that provides anxiety reducing significance to one's life, ethical codes of regulation, a theodicy of the distributions of fortune, and above all, they need to assuage the painful reality of the inevitability of death (Becker 1973). For most of human history, religion has provided people with meaning and direction for their lives, and in many cases, hope for some kind of immortality. Central to our thesis, transgression can provide compensatory identities. Following Weber, the 'disenchantment of the world' eroded the sacred and the realms of transcendent meanings-some would seek to 're-enchant the world' by flight from its shallow consumer culture.

We are suggesting that these moments of desire enable us to understand the affective basis of group life and social interaction. More specifically, we argue that people form groups not only for their instrumental or expressive functions of collective life, but group life allocates identities and social interactions that provide various emotional gratifications to the empirical subject. Otherwise said, membership in an identity granting community of meaning is the basis for much of our affective gratifications or frustrations.

Part III: Identity and Modernity

Fashion as Resistance

Castells (1997), argued that in modern, 'network societies', there are three patterns of identity, those that 1) *legitimate* the social order. But every system of domination fosters resistance if not rebellion. Thus we also note, 2) *resistance* based identities that may be progressive or reactionary and finally, *3) project* identities such as feminism, ecology, or global justice concerns that would transform subjectivity as well as the external world. But to his typology we would add a 4th) *ludic* identities, forms of selfhood that valorize play, privatized hedonism, and life style basedexpressions of selfhood. Ludic identities, systematically fostered by consumer capitalism with its mass proliferation of images of gratifying consumerism can take a variety of forms. As we noted, there has been a correlation of forces in which 1) adolescent rebellion becomes typically expressed in commodified forms, and 2) the culture industries promote hedonism and transgression as marketing tools. But 3) and perhaps most important, this creates spaces for cultural resistance against commodification itself identities of anti-consumption that, in the spirit of the Romantics, become a compensatory quest for 'authenticity' of self and experience. From what has been said, following Bakhtin, inversion, resistance and transgression may very well provide that resistance-in the encapsulated realms of urban subcultures wherein cultural transgression serves the sustain the political economy. Thus we see a number of subcultures of *resistance* expressed through the transgressive, vulgar grotesque and the carnivalesque in which ludic rejections and repudiations of the dominant culture serve hegemonic functions by fostering a migration of subjectivity away from political economy and neutralizing *resentment* by channeling it into privatized forms of cultural protest.

Following Nietzsche, marginal groups are likely to feel *ressentiment* toward the superiors, they invert and repudiate their values. These 'transvaluations of the ethical', valorize what is proscribed and celebrate what is disdained. Much as Merton described 'innovators' as those who would change both the goals and means to attain those goal, ludic identities of resistance at the margins of 'polite society', interrogate and resist the dominant culture to valorize other means and ends. As we have noted and will indicate, body art/modification, using oneself as a canvas of the grotesque to secure membership in the liminal anti-structures of inversion, gives meaningful identities and voices to those who might be invisible.

Transgression: Subcultures of Resistance and Liminal Identities

One of the long standing debates among urban sociologists has been the question as to whether or not urban life fostered weak, attenuated social ties or enabled liberation and freedom or more likely, both. But for urban subcultures, this is not so much a contradiction as a syncretic convergence, that is the weak ties and anonymity of the city, as well as its freedom and toleration, provide various subcultures with spaces where they can form highly cohesive, meaningful communities. (Hostility and/or disdain from the larger community also helps sustain solidarity). Youth, as a stage in the modern life cycle concerned with autonomy from parents and forging a separate identity, typically become members of youth cultures and/or subcultures that can be seen as identity granting communities of meaning which give the person a sense of belonging, provide him or her with recognition of his/her self, provide a sense of empowerment, and in general, assuage the 'ontological anxieties' of modern life Giddens (1992). Thus participation in the subcultures of youth, above all else grant dignified identities to those within the boundaries of the group. While this is important in all youth groups, to gain a positive sense of valued selfhood becomes *especially* important to groups of youth in marginal class positions without status based deference or promises of participation in the dominant society.

For a number of young urban people especially those 'surplus populations' now marginalized by the structural changes of globalization, (or pro-actively anticipating marginalization) there develop a variety of identity granting counter cultures that would reject the meanings and values of the dominant society. Among the most important signifiers of late modern youth are the various subcultures of transgression such as heavy metal, hip hop, grunge, punks, goths and ravers.etc. While these groups differ from each other, they share certain crucial qualities; they embody and celebrate the ludic, hedonistic and transgressive. These liminal anti-structures provide their members with meaningful, gratifying identities and social interactions. Among these countercultures and anti-cultures, there are a variety of distinctive styles of dress and bodily adornments from hairs styles/colors to make up to tattoos and piercings that valorize transgression and inversions of propriety that become the norm. The vulgar and grotesque become 'beautiful' as they empower and elevate the status of the bearer and lower the status and prestige of the 'uptight' main stream. Indeed the grotesque inverts usual categories of hierarchy. The grotesque is

> .... a powerful esthetic category that disrupts and distorts hierarchical or canonical assumptions. The notion combines ugliness and ornament, the bizarre and the ridiculous, the excessive and the unreal. The term derives from the Italian term for grottos, i.e., the ruins in which ornamental statues of distorted figures were found in the XV and XVIth centuries (*grotteschi*). The Romantic era, with its interest in individualism, and in all those who before the age of Revolution had been nameless and invisible, made the grotesque its indispensable adjunct...Bahktin placed the grotesque at the heart of the carnivalesque spirit. In the realm of the fantastic, it is a powerful weapon used in the revelation and denunciation of constructs. http://www.skidmore.edu/academics/fll/janzalon/grotesque.html

We would like to focus on two aspects of the grotesque as identity expressed in adornment and fashion.

Identity, Fashion and Adornment

In every society, people dress and adorn themselves according to social conventions and the realization of 2 desires, differentiation and inclusion (Simmel, 1950). Thus in a modern society, people seek some public displays of self articulation to differentiate themselves from others and at the same time foster inclusion into their own group. Historically, this has been base on categorical status, gender, occupation or other indication of social rank. But in modern society , fashion as style, has become an important component of personal identity and public statement of selfhood. One of the most important aspects of modernity was the importance of fashion in clothes and adornment as markers of self and group membership (Simmel, 1950).

But since the time Simmel was writing, there have been at least 3 changes in fashion, 1) At one time, fashions, whether in clothes, art or intellectual tastes, largely flowed from the more privileged and sophisticated classes downward to the masses and less educated. This has changed and a major aspect of current fashion has been to market the styles of the urban poor-typically African Americans or Hispanics set the standards for Caucasian and Oriental youth. 2) Secondly, while clothes may have been a marker of political orientation, eg white sheets, black shirts or work uniforms may have stated support for or resistance toward certain *political stances*, today, the body, clothes and forms of adornment have markers of *cultural resistance*, opposition to the dominant values and life styles that are expressed in personal style of life rather than political mobilization. The fashions can be seen as reinterpretations of conflicts of class/culture or both in the larger society. Fashions are a statements of who you are and who you are not, your ingroup and the outgroup, -and body modifications celebrate those boundaries. Sanders notes the tattoo as both an indication of disaffiliation from conventional society and as a symbolic affirmation of personal identity (Sanders 1988:395). 3) Moreover, the body itself has been turned into a 'work of art' in which agency confronts the normative disciplinary practices to recreate an 'aura' in an age of virtual reproduction. Thus as we shall argue, these themes come together in some of the more extreme fashions of youth cultures of resistance, can be understood as expressions of agency in which the body itself has become a medium in which liminal identities valorization of the grotesque, the bizarre and the vulgar can be articulated.

Part IV: Varieties of Modified Bodies

In terms of resistance as disdain and rejections of the dominant culture, a variety of public identity statements would cross the boundaries and repudiate the larger society-especially articulated in fashion and adornment. Wilson (1987) suggests that the 19th C. dandy was the archetype of anti-fashion-oppositional style hostile to the conformist majority. Such clothes were and designed to shock. But as we shall see, what is especially interesting about contemporary forms of fashion is the use of the body and adornment of the body as fashion. The body, long the center of the private realm, has itself become the template up which aesthetic sensibilities are inscribed.   Perhaps inversion, indeed transvaluation, is nowhere more evident that in the decorations and modifications of the genitals-what had been made private and hidden through shame, have become adornments and bases of pride. While not typically displayed in 'polite' company, the various rings and things are proudly shown off within the alternative community.

For Simmel, the original basis of covering the genitals was adornment to call attention to the sexual parts. Only later did shame emerge as a means of social control and with it, people knew they were naked and felt ashamed-so they covered their private parts with the leaves of figs. Following his insights, we would note how for certain cultures of resistance, making the private public, especially the pubic, can be seen as a rejection of Western civilization and its repressive morals, values and its constraints. 'What seems evident is that in traditional societies, ritual body modification practices connect people and their bodies to the reproduction of long established social positions whereas in the industrialized West body piercing seems to serve the function of individuating the self from society. (Holtham, 1992) In her study, she suggested that:

> Whereas Bill de-emphasizes the sexual aspect of body piercing and says his involvement is to do with 'the mapping of his own history and personal evolution'. The many piercings and tattoos he wears having a 'synergistic effect on his life', facilitating an 'alternative spirituality'. Bill and George emphasize what they call the 'tribal aesthetic' in both their own bodies and in their piercing studio in an effort to 'honor the tribal roots of body piercing'. George's interest in piercing stems from what he describes as 'the desire to experience his mind and body on different levels and thus become more self-aware.

Louise describes the demand for professional piercing services in

Melbourne as having 'gone through the roof' in the last three years. Her business began as a two hour service on a Saturday afternoon above a gay male adult bookstore in 1990 and has since expanded to two studios in Melbourne and Sydney, with six piercing staff working full-time six days a week. Louise credits the phenomenal rise in interest in body piercing to a range of factors. Predominantly the increased visibility of piercing having a kind of 'flow-on' effect as new people become aware of the options available to them to augment their bodies. Louise also performs many piercings as part of commitment ceremonies and ritual acts of submission or attachment. She also senses in her clientele a general move toward 'reclaiming their bodies and taking pleasure in the look and feel of the piercing'. The three interviewees all credit the rise of interest in body piercing in part to do with a general disaffection with the governing narratives of our lives. Specifically, Louise thinks many of her clients are 'on the search, searching for meaning, and for feelings of belonging to something larger than themselves that isn't religious' (Holtham, 1992 p.7) Thus we can see that body modifications become an important element of transgressive identities that provides the person with profound lived experiences, incorporation into a life-style of resistance that provides solidarity and meaning in a world where both are ever more problematic.

Urban Primitives

Interestingly, in the English language, the word 'primitive' was first used by the Christian Church to refer to the pure beginning of its own religion (' the primityve churche of Christ'). Over time, its meaning slightly changed and was used to describe that which was original in form, as opposed to a reproduction. In the 20th century, that the term acquired a negative connotation, it became used to describe non-Western appearances and activities, the strange, bizarre ways and appearances of 'uncivilized' people, read: without shame over their bodies and sexually free. Western Europe and Euro-America espoused the *Ideology of Progress*, where heterogeneity threatened humanity's unitary ascension to greater knowledge, sophistication, and 'goodness.' 'Primitive' became a powerful label used to describe the ignorant, barbaric, and uncivilized (Strauss, in Vale and Juno, 1989). Cultural differences marked the stages of development, with the West representing the most developed and advanced of all. Cultural diversity was to be stamped out and replaced by refinement, enlightenment, 'civilized behavior.'

In Western culture, forms of body modification such tattoos, scarification, and piercings have typically been regarded as a regression toward primitive practices. What's more, it is in the Judeo-Christian context that tattooing and modifying the body is regarded as transgressive. *Leviticus 19:28*: 'You shall not make any cuttings in your flesh on account of the dead or tattoo any marks upon you.' Yet, in other non-Western cultures, body modifications are considered 'sacred or magical, and always *social*. The unmarked body is a raw, inarticulate, mute body. It is only when the body acquires the 'marks of civilization' that it begins to communicate and becomes an active part of the social body' (Strauss, in Vale and Juno, 1989). While tattoos, piercings, and other forms of extreme body modification are not considered 'civilized' modes of communication in mainstream Western culture, there are indeed social. By transforming the body into a medium of identification and art, one is actively defining and exhibiting a liminal identity located in a subculture of transgression. Notwithstanding the repulsion by the larger society, indeed as both a resistance against that society and the spaces it nevertheless creates, such identities nevertheless locate and connect the person with others, provide recognition, empowerment, and a framework of meaning. The gaze of the Other does not turn one into an object without freedom as Sartre would have it, but by commanding the Other's look, renders him/her the object.

Facing the World

Hair is one of the first noted aspects of physiognomy and accordingly, coiffure reflects cultural tastes-and distastes. In 'primitive societies' where little clothes were worn, hair and tattoos were typically markers of identity and inclusion. In the biker cultures of the 50's the DA, duck's ass, was *de rigueur* for all rebellious teenagers, and at the time, was a very daring rejection of dominant norms. One of the defining markers of the counterculture of the 60's was 'hair, hair…lots of hair' as the musical proclaimed. As a rejection to the trim crew cuts of corporate life and/or the military, long hair, afros etc, became the initial markers of membership in subcultures of resistance that would defy social, political or cultural conventions. It was at this time that hair styles became important for the images of popular musicians, the Beatles initiating the radical styles of the early sixties, while till this day, most male musicians have long hair.[1] We might also mention how beards again became fashionable, at that another

---

[1] Classical musicians have long been referred to a long hairs-marker of both elite tastes and an insouciance to the standards of the philistines. Thus when popular music moved from the bottom up, its performers appropriated, in exaggerated form, the styles of elite performers.

rejection of the 'clean shaven look'. Further more, the beard might designate wisdom and masculinity.

For our purposes, what become important are the more extremes of coiffure, multi colored, multi spiked, teased, locked etc. From what we have said, hair stands as one of the most blatant markers of fashion and identity, it is easily recognized. Following musicians and the long hair of the 60's counterculture, among the punk cultures that emerged in the 80's, the Mohawk became an important marker of working class resentment, soon appropriated by others seeking to articulate resistance. Perhaps De Niro's character in *Taxi* made this point. From this followed various spikings and dying/tinting hair in colors not found in nature, bright reds, yellows, greens, blues and often in phosphorescent hues. That three spikes-in three colors- stands out as a marker of difference is easily gainsaid. More recently, braided hair/dreadlocks have become popular among those who would identify with Jamaican culture, often identified with pot and Marley that has come to represent another version of the 'primitive', at least a simulation of 'primitive' spontaneity, Dionysian excess and vitality-the very opposite of the 'civilized' Westerner. Finally, for men, the pony tail, a maker of androgyny, is often found among artists, people in computer technologies, advertising industries-and a few academic specialties like communication, cultural studies and theater. (And even in sociology there are men with pony tails) As an indication of resistance, that can easily be trimmed and/or dyed back to the natural condition, hair styles, much like henna tattoos, need not be permanent.

Tattoos

The tattoo The tattoo may well be the prototypical marker of the

'primitive' and indeed, early travelers to Polynesia, Samoa, New Zealand (Maori) and Australia, including Melville, often remarked on the elaborate body designs of the native peoples Although bodily enhancement may be relegated to specific cultural norms, within those limits an individual has a great amount of room to use decoration as an expression of their own personality and what they conceive as their 'self'.

The [Maori] tattoos identified who the person bearing them was within the society....the person's position in life, his or her rank and parentage, their marriage status, what line of work they were in, and their 'mana' or power bestowed by the gods (Vale and Juno, 114). Facial tattoos were used as signatures when Europeans became involved, demonstrating the personal link between the design and the individual. Tattoos showed political allegiances and tribal connections. The only people who were not tattooed were sacred virgins and rulers, those who were outside the scope of everyday social interaction. This lack of tattoo identified these individuals as different and 'special' (Wall, 1998).

The tattoo has had a long history as a proletarian art form, initially embraced by seafarers and adventurers who came in contact with the natives of the South Pacific. Eventually it was embraced by and urban workers where it often became a visible marker of class difference,realmen, workers, had tattoos as opposed to the higher status effetes of the offices. As an urban art form of workers, prisoner, bikers etc, mainstream society began to regard the tattoo as the mark of an unsavory character. Today, tattoos have become appropriate among large number of youth-in part because of their unsavory connotations. Tats are considered, cool, artistic, hip, counter culture, they have become adornments that are adored. Moreover, they are as likely to be found on women as men-an homage to the imagined equality of primitive society. Certain tattoos have an important erotic function, especially the relatively small butterfly, flower or design that some people, mostly women, place on the butt, pubis or inner thigh. The function here is to create a mark of intimacy, known only to the lover(s). Finally, it might be noted that one of the most popular forms of tattoos, often found in beach based tourist destination as well as cities is the henna tattoo which unlike the engraved inks, will fade away in a few weeks.

Rings, Posts and Studs

The Nose

The rejection of modernity, the embrace of the primitive is most clearly articulated when stainless steel, silver, titanium or titanium replace bamboo, wood or bone for piercings. The representation of the 'primitive other' with the bone in his/her nose has long been a part of defining the modern in relation to the primitive 'Other', the ascetic and unadorned was 'good' in relation to the 'evil' impulsive, uncivilized 'savage'. The ringed nose in your face-as well as the face of the piercee is the expression of transforming passivity to activity, through the disdain of repression and marker of the rejection of the modern. As befits inversion and the vulgar, the ring is always the reminder of the flowing of snot, another repulsive fluid of the body. Moreover, as earlier noted, and reminiscent of Freud's early relationship with Fliess, the nose is the intrusive phallus, as in keep you nose out of my business. In the hierarchy of facial adornment, the septum ring, the bone it the nose, stands at the starting point. Surely, the many others, the nostril pin, stud or ring, eyebrow rings, posts and tattoos join the chorus, but rarely as stars. Thus many women might have a small stud in a nostril, much like traditional Indian women. Others opt for small nasal rings.

Satanic Horns

Perhaps the ultimate symbol of resistance and repudiation of the dominant culture and its notion of the 'good, is the celebration of Satan and the embrace of evil. In the inverted world of the grotesque as beauty, where there has been a transvaluation of the ethical, the ultimate evil becomes the ultimate good, the fallen angel becomes the simulated *ubermench*. And thus we that the implanted horns that advertise inversion to all those who would enjoy the sceptic pleasures of the liminal. But even more important, the horned person commands the gaze of the other and in commanding that gaze, making the 'normal' Other view those horns, making him/her repulsed by the repulsive, the horned 'devil' triumphs over *das Mann*.

Beware of Forked Tongues

In the Judeo-Christian West, the Fall of Humanity begins with the snake, the slivering phallus that embodies pure evil. Satan after all was a fallen angel and might have a trace of residual goodness, while the snake is without any redeeming value. In Western popular culture, the snake is often the 'ur' symbol of evil, the floor of the Temple of Doom is covered with snakes. Former WWF wrestler, Jake the Snake was the personification of the baddie who let his snakes play on his defeated adversaries. We yet warn of speaking with forked tongues. This the split tongue of the snake or lizard, much like horns, celebrates what had been evil and indeed, much of what we said about horns can be said of forked tongues.

Pubic Art: Ring and Things in One's Things

As we have suggested, one of the most important moments of the rejection of modern asceticism and denial of embodied, erotic desire is the valorization the genital and inverting the shame of expose of one's genital that are typically private into the pride based public. We might call this pubic art, decorations/piercings of one's genitalia and/or proximate parts for purposes of display to others. Such displays are discreet in that as part of a subculture, are typically only displayed within that subculture. It is interesting that one of the more prevalent form of display has become the Internet and a number of sites, (http://www.bme.freeq.com), require a photo of one's piercings to be posted as the entry requirement to the members only sections.

There are a large number of various rings, studs, posts and other things that people do to their genitals, the most common of which are rings. Perhaps the one known most widely is the Prince Albert, a single ring in the glans, through the urethra, purportedly used to tie it to the leg. But men often have several rings, locks and other things implanted into their penis, scrotum and perineum. Some prefer posts to rings and a series of parallel posts is called a ' ladder'. Similarly, women are likely to put rings in their nipples, labia, clitoris, hood etc. Quite often women report that such piercings enhance their sexual relations and are often provide sexually stimulation in everyday life. (We are reminded how in the days of foot operated sewing machines, occasionally some of the seamstresses would begin to sew at much higher rates-at least for a short time-and utter a load sigh of relieve as they finished the cuff). Moreover, unlike more repressed generations of the past, in any body modification chat groups (as well as S+M) crowds, women will freely and openly discuss their sexuality in ways that would and do shock the dominant culture.

Big Genitals = Much Power

We have argued that the embrace of the primitive stands as an inversion that empowers, that one reclaims agency by transforming the body. Just as grotesque may be exaggeration, so too can and do body mods use saline injections to mold and shape, and most of all amplify through swelling certain parts of the body. For some 'modern primitives', when it comes to genitals, bigger is better, size does matter. For men, the notion that size matters always confronts physics when many try to stretch their penis. As Newton showed, there is a law of the conservation of penile matter, the longer its stretched, the thinner it gets, till we have the 24' noodle. But there is a quick and simple alternative, the saline injections into the scrotum. In a matter of minutes, one has 'big balls', the universal symbol of male power-another indication of reclaiming activity from passivity.

Similarly, in the world made egalitarian by feminism, or is it perhaps the reclamation of the egalitarian world of the pre-modern primitive that existed before the 'Motheright' was overturned, the woman is again powerful. In much of the statuary of pre-modern South America, Africa and Polynesia, females are often seen with highly exaggerated vulvas-celebrations of her fecundity and role in reproduction. Thus we see that if a hypo of salt water can give guys big balls, so too can a hypo can give the woman exaggerated labia, mons or both. If an inflated scrotum can give him symbolic power, inflated labia can restore equality and perhaps indeed express the empowered woman.

Conclusion

A dialectical understanding of history shows how societies contain within themselves their own negations. For Marx, capitalist political economy led to alienation, crises of accumulation that would lead to the overcoming of capital by democratic socialism. Much the same argument might be said about culture-especially since we might have to wait awhile for a globalized socialism. We could argue that the repressive values of modernity regard anality as normality. Christian repression endures but without transcendental meaning, yet has within it, a historical basis of its negation. More specifically, below the surface of the modern Apollonian, rests   the earlier Dionysian of the Christian carnival. Like Godzilla awakened from the depths of prehistory by modern atomic weapons, so too have the urban cultures of modernity revived the long submerged grotesque of their own history. More specifically, capitalism in its consumerist form required the erosion of its underpinnings of restraint and control. Thus advertising inspired consumerism colonized consciousness to unfetter desire to then be realized in shopping and mass culture. Media, especially film and television, have encouraged the erosion of standards of taste such that we have seen the return of the carnivale in commodified form.   At the same time, advanced capitalism has transformed itself into a vastly profitable globalized profit machine that is rapidly changing. To achieve obscene profits, capitalism has employed leading edge technologies, computerization, job export and deskilling, and delayering of management in favor of quality circles that have had a major impact on the labor forces of the advance industrial societies. There has been an ever smaller number of people who amass great wealth while the majority of people, the ones in lesser skilled service occupations, have been steadily losing ground. For many such people neither the available jobs nor the dominant cultures of modernity hold much lure for defining one's self. In face of these transformations, there has been a wide range of responses. Some would find meaning in religion, at least conservative religions that provide stability in a sea of change. Thus many have embraced fundamentalisms or cultural conservatisms. But many people, especially those marginalized by the economy, have moved to the limits of the cultural boundaries that were themselves been rapidly shifting as the culture industries have spearheaded the carnivalization of culture and celebration of the grotesque as entertainment. Once repressive culture permitted, indeed celebrated the expressions of desire, like opening the crypt in the horror movie, the unanticipated emerged. Thus we have seen the proliferation of a number of urban subcultures of resistance, transgression and inversion that provide people with compensatory alternatives that valorize identities of transgression, inversion and resistance-and in turn gain attachments, recognition, empowerment and meaning. But joining certain subcultures often requires dramatic initiation ritual and/or various accouterments or markers of having moved from the dominant society to subcultures of opposition and/or resistance. These various moments come together in the subcultures of body modification- and insofar as many of its proponents see themselves as 'modern primitives' reclaiming their bodies through primitive markings/piercings.

This 'lifestyle', or more importantly as part of what they regard as a movement toward what has been described as 'modern primitivism' is the harking back to something more 'basic' and fundamental in human nature: pain, ritual, a concern with the processes of the body and exploration of different levels of consciousness and physical experience. (See Juno and Vale, 1979). It is upon this common interest in adorning and modifying the body that small affiliational communities - or tribes - of 'modern primitives' have developed in the wealthy, industrialized Western nations. Holtham, 1992, p12. Certain anthropological insights are relevant. In many age graded societies, dramatic 'rites of passage' mark the transition from one stage to the next (Van Gennep, 1961). Myer's participant observation in ritualized Many of these ritual involve various painful bodily modifications such as scarification, branding, tatoos, or genital mutilations (circumcisions, subcisions, various vaginal cuttings now generically considered FGM). Sado-Masochistic group 'piercing parties' over a two year period led him to conclude that such practices fulfilled a universal human function in providing a 'rite of passage as a cultural drama' as well as providing the means by which members could proclaim their various social affinities (Holtham, 1992) Insofar as the modifications indicate transition to a higher status, they are eagerly sough by the initiate and celebrated by the community. Parenthetically, psychological research has suggested that if initiation rites are painful and dramatic, the groups are more highly rated and evaluated-any group whose entry requires much suffering must surely be worth entering. In the present case, body modification is not only the means of entry into certain subcultures, but as a public, visible marker, a matter of fashion and identity.

With Simmel (1950) we see the function of vanity, wherein the adorned individual accentuates his personality by wearing jewels and metals, which simultaneously invokes both admiration and disdain from others. He distinguishes and amplifies himself in the view of and at the cost of others. To be extraordinary, however, is not simply a matter of exhibiting the material objects themselves; but further, a display of it's 'more-than-appearance value.' Simmel, 'it is not something isolated; it has roots in a soil that lies beyond its mere appearance, while the unauthentic is only what it can be taken at for the moment' (p. 342). Thus, with regard to the extreme piercing and dress, it is not only a 'fuck you' to mainstream society, but also a method of generating a reaction of disgust and fear. It is the antithesis of the admiration generated from mainstream fashion and couture. After all, one who would vacuum, reroute, and inject saline and posts into his or her most sensitive body parts, clearly does not fear pain. Consequently, to adorn and modify the body is a choice to not only reject mainstream conventions, but also to reclaim agency. The pain and identity (as liminal) is chosen, and the fashion serves as a statement of who the individual is and who he is not. Further, those modifications and styles that are clearly visible, call attention to one and thus almost mandate that one is looked at. But whereas for Sartre, to be looked at was to be rendered powerless, to be made an object, for the modern primitive, to be looked at, to *command* the gaze, is to have power over the Other and indeed, whether the gaze is the disgust by the mainstream, or the acceptance by the subculture, to be viewed, and to be viewed as 'different', is to gain empowering recognition. As most sociological accounts of life in the present age note, all that is solid, melts into air. The traditional sources of stability in people's lives, community, workplace and even family have been eroded. In an unstable world, extreme body modification becomes an act of identification and empowerment, where one is able to reclaim his/her body in a way that cannot be moderated by others and in ways that are relatively permanent. Thus, to accentuate one's personality, in essence to become extraordinary, allows one to take control of at least part of his existence and impose stability. While one cannot make certain that s/he will not be abandoned by a lover, or receive the dreaded pink slip at the monotonous desk job, there is a sense of permanency in being a 'body mod.' For even after the saline is drained, and the posts and rings are removed, the individual has become a sort of artist and transformed his body into a piece of art. Moreover, for those who take the modifications to the extreme of extremes, reversing the procedures, and fading back into the mainstream and pushed to the bottom becomes nearly impossible.

Walter Benjamin argued that art, when mechanically reproduced lost its 'aura', its unique qualities imparted by the artist. He saw that mechanical reproduction enabled a democratization of culture and pluralization of interpretation.   Jessica Benjamin (1992), in her psychoanalytic approach to female masochism suggested that women endure pain and humiliation to gain recognition. In her analysis of *The Story of O*, she described how the heroine 'voluntarily' submitted to sexual servitude and degradation to gain recognition and personhood. While the story is fictional, Benjamin suggests the story speaks to the real issues that many women face. If we join Walter Benjamin with Jessica Benjamin, we might note that to decorate one's body is to claim authorship of one's self and reclaim an embodied 'aura'. But insofar as this may well involve pain, that pain is an expression of agency and means to recognition and inclusion. (We would also note that for many women, genital piercings seem to both cause sexual pleasures and enhance the pleasures of sexual intercourse.) Further, as we have argued, insofar as global capital has displaced many from the contracting mainstream of the culture, the pain of modification becomes a way of authoring ones one pain, that seeking pain rather than being pained, turns passivity into activity, controlling pain and the impression that marking make on Others.

Postmodernists argue that we are now in an age without grand narratives, a moment when the modern social order has imploded into a plurality of subcultures and decentered subjects. While they have the description right, the analysis is quite lacking. Rather we argue that capitalism, in its global form, remains the grand narrative of our times. Similarly, the fragmentation of social was rooted in the class structure, and industrial division of labor, but flourished with the proliferation of segmented marketing producing a pluralization of life worlds of consumer subcultures and life style enclaves. Capitalism   then fostered the *carnivalization of culture* as a marketing tool to erode any restraints upon spending money. But at the same, globalization has produced a plethora of goods for these many subcultures, it has also produced a surplus of people whose job prospects are bleak and promises of stability are meager. These factors dispose a prolongation of youth, sociological understood as an extended moment of the life cycle beyond adolescence, but not fully integrated into adulthood, at least as it has been traditionally understood.

Thus we might also note the contradictions between the Appollonian nature of the traditional culture of control and restraint, and its negation in the Dionysian frenzy of passion and indulgence-much of which has encouraged by the very culture industries of late capital.

These forces have converged in the production of a plurality of sub cultures of resistance. To be sure, most of these subcultures valorize cultural inversions and celebrate the grotesque   rather than assume political forms that would transform the nature of globalization.[2] Among these many articulations of resistance, typically embraced by the young and young at heart, are the many expressions of body modifications that would reject modernity and embrace the primitive. As such, a proliferation of identity granting communities of meaning have emerged to provide people with the social attachments, dignity, agency and meaning that have become so problematic in late modernity. While many of us may well find their life styles, values and even their modified bodies grotesque, it may well be that their life style proves more adaptable to the rapid changing world of today-by escaping to a simpler, and perhaps gentler past, and surely one far kinder to the environment.

In the early part of the 20th C., as industrialization changed the face of Europe and America, sociology emerged to chart the rise of capitalism, industrialization and urbanism as a way of life. While Marx, Weber and Durkheim painted the broad outlines of the rise of rational capitalism in its industrial moment, with its advanced division of labor, it would remain for observers like Simmel, Park, Burgess and Worth to illuminate the details of urban life with all its complex and contradictory forms of social life. The city was the seat of rationality and tolerance, but at the same time, a collection of small villages or 'moral communities' where people of common heritage, class or life style might find peace and comfort with each other apart from the larger, more anonymous city. Further, within those various areas might be found communities of resistance and inversion, the 'hobohemias, the red light districts, the 'zones of transition' prone to crime, drink and unmarried sexuality. As Mills noted in his critique of the urban pathologists, sex in the city, shocking such things went on. Nevertheless, the logic of urban life endures in the modern city, the 'fantasy city' in which globalization and consumerism have converged. But in face of these enormous changes, so too do we see both certain continuities of urban life as well as distinctive features that speak to the present age.

---

[2] The alternative globalization movements can be seen as the political expressions of these same factors. See Langman, Morris, Zalewski, 2002.

The city then and now erodes social ties as well as creates spaces for and toleration for the 'deviant' communities fostered by the larger social order. The city with it large populations insures that even when a certain lifestyles and values are rare in the population, there will be more concentrated numbers in the city. As has been argued, one of the most distinctive expressions of the responses to globalization, and its transformation of the opportunity structure has been the growing popularity of 'urban primitives' characterized by body modification and adornment as urban ways of life-or perhaps we should say rejections of urban ways of life. By decided choice and often great personal pain, a growing number of people elect identity granting subcultures of meaning that repudiate the dominant codes of repression and 'propriety', of consumerism and anality, and instead valorize the vulgar, celebrate the grotesque and act out the inversions. Yet as we recoil from our shock at how such people transform their bodies, we must note that their life styles are telling us a story about our age. Hopefully, we have not only told some of this tale, but note that their life styles may be much like the canaries in the mine that warn of dangers long before people are affected. So too may the urban primitives may be offering a more trenchant critique of globalization than do sociologists.

References

Bahktin, Mikhail. M. 1968. *Rabelais and His World.* Cambridge, MA: M.I.T. Press.

Becker, Ernst. 1973. *The Denial of Death.* New York: Free Press.

Benjamin, Jessica. 1992. *Bonds of Love: Psychoanalysis, Feminism and the Problem of Domination.* New York: Pantheon.

Campbell, Colin. 1987. *The Romantic Ethic and the Spirit on Modern Consumerism.* Oxford: Blackwell.

Castells, Manuel. 1997. *The Network Society,* Vol.II: The Power of Identity. Malden MA. Oxford UK: Blackwell Publishers

Clark, Terry N. 2004. The City as an Entertainment Machine. *Research in Urban Policy 9.* Amsterdam: JAI Press.

Elias, Norbert. 1978. *The Civilizing Process.* New York: Pantheon.

Florida. Richard. 2002. *The Rise of the Creative Class: and how it's Transforming Work, Leisure, Community and Everyday Life.* New York: Basic Books.

Freud, Sigmund. 1989. *Civilization and Its Discontents.* New York: Norton and Norton.

Foucault, Michel. 1977. *Discipline and Punish: The Birth of the Prison.* New York: Pantheon.

Giddens, Anthony. 1991. Modernity and self-identity. Self and society in the late modern age. Cambridge (Polity Press)

Hannigan, John. 1998. *Fantasy City.* London: Routledge.

Holtham, Susan. 1992. 'Body Piercing in the West: A Sociological Inquiry,' *Http://www.bmezine.com/*

Honneth, Axel. 1995. The Struggle for Recognition: The Moral Grammar of Social Conflicts. Cambridge: Polity Press..

Korten, David. 2001. *When corporations rule the world.* San Francisco: Berrett-Koehler Publishers.

Langman, Lauren. 1992. 'Neon Cages: Shopping for Subjectivity,' in Rob Shields, Ed. *Lifestyles of Consumption.* London: Routledge.

—. 2000. 'Identity, Hegemony and the Reproduction of Domination,' in *Marx, Weber and Durkheim.* New York: Gordian Knot Press.

DeMello, Margo. 2000. *Bodies of Inscription: A Cultural History of the Modern Tattoo Community.* Durnham: Duke University Press.

Perucci Robert and Earl Wysong. 2003. *The new class society: goodbye American dream.* Lanham: Rowman & Littlefield.

Postman, Neil. 1986. Amusing Ourselves to Death: Public Discourse in the Age of Showbusiness. New York: Penguin Books.

Ritzer, George. 2000. Enchanting a Disenchanted World: Revolutionizing the Means of Consumption. Thousand Oaks: Pine Forge Press.

Sanders, Clifton.1988. Marks of Mischief: Becoming and Being Tattooed. *Journal of Contemporary Ethnography* 16.4: 395-432.

Sassen, Saskia. 1998. *Cities in a World Economy.* Thousand Oak: Pine Forge Press.

Schilling, Chris. 1993. *The Body and Social Theory (Theory, Culture and Society Series).* London: Sage.

Sennet, Richard. *The Corrosion of Character.* New York: W.W. Norton & Company.

Sennet Richard and Jonathan Cobb. 1972. *The Hidden Injuries of Class.* New York: Knopf.

Simmel, Georg. 1950. *The Sociology of Georg Simmel.* Translated by Kurt H. Wolff. New York: The Free Press.

Skalir, Leslie. 2001. *The Transnational Capitalist Class.* London: Blackwell.

Stauth, Georg and Bryan Turner. 1988. *Nietzsche's Dance: Resentment, Reciprocity and Resistance in Social Life.* London: Blackwell.

Stallybrass, Peter and Allon White. 1986. *The Politics and Poetics of Transgression.* London: Methuen.

Taylor, Charles. 1992. *Multicultur and Society*. London: Sage,
Turner, Brian, 1996. *The Body and Society*. London: Sage,
Turner, Victor. 1966. *The Ritual Process: Structure and Antistructure*. Chicago: Aldine Publishing Company.
Van Gennep, Arnold. 1961. The Rites of Passage. Chicago: University of Chicago Press.
Vale and Juno. 1989. *Modern Primitives*. San Francisco: Re/search publications.
Wall, T. 1998. *Decoration of the Body and Self Expression: A Theory*. *http://hamp.hampshire.edu/~tawf95/bodmod.html*
Wilson, Elizabeth. 1987. *Adorned in Dreams*

NETWORKING AS DISEMBODIED CAPABILITY BUILDING: TESTING SEN'S CAPABILITIES

CARLO DONOLO

The text is based on the paper presented at the XVI ISA World Congress, Durban, South Africa, WG3 'The Body in the Social Sciences', July 2006. I am grateful to Professors Roberto Cipriani and Biancamaria Pirani for their comments and suggestions.

Abstract: The notion of person – the social persona ficta produced by different normative instances – will be tested, considering its roles and functions in the networking space. Social networks and virtual networks melt in ordinary experience and also in the different cités of postmodern social life. The networked actor will be very different from the atomistic, autocentered subject presumed by normative approaches offered in the philosophy of law, the rational action approach, and even in policy making. Starting from the socializing impacts of networks, I will search for their enabling effects, especially in the perspective of the capabilities-functionings frame developed by A. Sen and M. Nussbaum. Networking produces virtual worlds where the actors – more and more personae fictae – become disembodied or virtualised. Virtuality and disembodiment in capability building are delineated as a path of paideia for the hypermodern social person.

Keywords: Body, person, network, capability, virtual world

1. Some General Assumptions Concerning Persons, Networks and Bodies

We start with the following assumptions:

a. Persons and every other sort of actors are built with and through institutionalizing processes. Such processes elaborate social matters (social dilemmas, normative stances, conflicts) confronting them with given institutions and opening new avenues of meaning and sense.

b.   Actors are social constructs, and as embodiments of social relationships, they are *personae fictae*. In such a sense, they lose partially their physical connotation to acquire the form of a social body. Social norms lead us to recognize such bodies – through a trial and error process guided by social standards –as essentially by-products of social processes. Recursively, only socially constituted bodies are able to identify and to accept/refute the normative component of the constituting process itself.

c.   Actors are more and more submerged in a networked space. Social networks and virtual networks melt in ordinary experience and also in the different *cités* (following the terminology introduced by Boltansky and Chiapello, 1999) of postmodern social life. The networked actor will be very different from the atomistic, autocentered subject presumed by normative approaches offered in the philosophy of law, in the rational choice approach, even in economics and in policymaking. At the end, we have to consider primarily networked and networking personae fictae as virtualised bodies and selves.

d.   Networks are virtual worlds, producing - by their very operation - multiple virtual worlds, each of them postulating or requiring an *apt* social actor, that is to say an actor socialized through the network, or the work-in-the-net. We have yet to fully appreciate the difference between being socialized through historically formed processes and institutions and being socialized through the networking experience. It is too early to fully recognize the differential impact of networks on social life, but it is certain that futurible impacts are already firmly settled within us, tacitly or explosively (Castells, 1996).

2. Understanding Networks as Holograms

Networks live in a world of diversity (Ostrom, 2004), that they unremittingly co-produce and transform. There are social networks going from the local to global range; they can be formal or informal, treating and handling money or Anerkennung with the same agility. There are institutional networks, streaming down the destructured Leviathan, where new protagonists do affirm themselves: cities, regions, territories, profit and non- profit organizations, autonomies, social movements, countervailing powers. There are the economy networks, based on productive districts, transregional fluxes, shifting between real and financial businesses, coalescing between localistic bonds and WTO rule.

There are friendly and deadly networks, networks for good and networks for doing bad, some are absolutely necessary to our lives, others are redundant and futile; many networks are wanted and others that are very needed do not yet exist. We also have networks of networks, and perhaps more invisible than visibile networks. In any case, the network is always with us, salient in promises and threats. We always know the net as a *chose sociale*. Thinking about networks, we have to hold together the social and the virtual dimension, the mental and the tecnological, the human conversation in natural language and the digital stream of signs and data. Surely the main trend is towards the growth of the artifiical and virtual component, with a corresponding contraction of the social and material factor - as a Mondrian hung in front of a Rubens. The different dimensions do not collapse and annichilate, but time, space, social dimensions, and ecosystems reproduce themselves in second, tirth and infinite virtual worlds, as a *mise en âbime* of reality and experience. Before the setting up of a digital world, we had already encountered that deepening and widening of experience through the performances and works of the artistic avantgardes, musical or visual.

Holding together the several dimensions, each with its distinct space-time relationships, we may think of networks as instituted, self-instituing holograms, i. e. virtual objects, built up with cognition and deep emotional affects, stretching and spreading out in different dimensions: physical, techonological, social, organizational, systemic, and normative. To *see* networks and to value their disemboding impact on persons we really do need the most developed 'conscience of the eye', as Richard Sennett suggested (1990).

3. Networks as Commons

Networks are commons. There are many types of commons and of networks. Single, local networks and commons bind together to build up a great chain of being, going up to Gaja, or to the global system. Networks are specifically apt at operating as intermediate steps in these chains and layers, which are so often recursive and autopoietic. We know the natural commons, with their articulate phenomenology and terminology, very well. We have also learned to recognize artificial and virtual commons: languages, encyclopedies of knowledge, motives, icons, norms, Lebenswelten, ecologies of social games, Idealtypen, imagined communities and so on. All are resources, media and also ends in themselves. Natural commons are hyperbolically multiplied in the artificial and virtual dimension through innovation, ibridization, communication, and dispersion. Natural and artificial commons are intrinsically intertwined, as when we say that organizations are machines, but also symbolic worlds, interactions, populations, and cultures. Organizations are artificial, but they transform themselves quickly into quasi-natural ecosystems, ruled by causality, path-dependency, automatic routines, and therefore, exposed to breakdowns, bricolage and re-engineering (White, 2002). The social and virtual commons are often tricky, ambivalent, scarce or devalued, having the tendency to build up inacceptable or impossible to understand situations for actors. Commons have to be governed to avoid the tragedy of commons; that statement applies in greater measure to networks as commons. Networks, as compounds of socially produced commons, have a preference for self-regulation, following the governance model of common pool resources (Ostrom, 1990), or require forms of governance scaled on plural levels, especially indirect rule through repution, moral suasion, incentives, ostracism, independent authorities, arbitrages...). Otherwise, they are exposed to the risk of being subdued by financial or technological, strong powers, and of losing their non-hierarchical form (Teubner, 1993; 2004). As a defence from such dominant strategies, networks often react through the enactment of autonomous actors from below and from the periphery; in that case, the social actors re-interpret their persona ficta status and regain autonomy from the resources of virtual worlds (Perulli, 2001; Beck, 2002) . In the end, it is not a surprise to recognize that the governance of networks can be only a net of networks, where governed and governing actors do not so much coincide (as in an ideal direct democracy), as become indistinct (Barabasi, 2004).

We need every sort of commons, and we aspire to become capable of reproducing and enhancing or respecting any common we need. Commons, especially at the intersection of natural, cultural and virtual commons, are enabling resources and contexts. Networks are assuming a central role both as commons and as the medium of virtuality. Thus, commons and networks become the ladder for any paideia of social persons, for the individuation processes wherein virtually disembodied personae fictae and bodies are searching for a better self.

4. Towards Virtual Worlds

Now let us develop some of the arguments following these premises and summary reflections. One of the relevant differences concerns the status of the body itself. Also personae fictae, such as formal organizations have a social body (an identity, a material logistics, and consume time and space in a very naturalistic sense). Bodies by definition, in social contexts, are both natural and social, or physically and symbolically constructed. Metabodies, such as institutions and organizations (we often speak metaphorically of social bodies to mean such aggregations), are consuming organs – in the terminology of Thomas Schelling (1978, 1984), and their communicating activities spend energy and other socially scarse and valuable resources. So, their materiality is self-evident. This will be easily recognized by every form of cultural materialism. With the growing impact of the knowledge and information society, the social body tends to acquire an increasingly virtual dimension (Taylor, 2003). The consequence will be to emphasize or exasperate the virtuality inherent to every symbolically constituted social world from the beginning. Virtual worlds do not lose their connection with material presuppositions such as energy, organization, the efficient use of resources and so on. But they are able to connect energy and information in more performative ways so as to lengthen and lighten the chains of such connections. In this way, they acquire new grades of freedom and autonomy, as theories of complexity suggest. Such qualities are revealed primarily through the primacy of self-regulating and autopoietic processes in the virtual world. The same can be stated for social bodies, that are primarily personae fictae (i.e. not existing outside of institutional and symbolical vocabularies of motives or *justifications* – Burke, 1962; Boltansky and Thévenot, 1991). Their first grade virtuality, as a product of history and institutions, acquires a second level virtuality through their navigation in the networked world. Such reinforcement transforms the very idea of body in its relationship with the idea of social person. Persons, having lost bodies as personae fictae, are now losing the space-time identificability. In fact, actors become knots. Knots have no body, but the same eco-localization is only virtual, or workable only in a virtual informational time-space. We can speak at the end, and in this context, of disembodied actors. We have to be careful about such a statement. We have to always keep in mind the material presuppositions, that are impressive also for the global network, in terms of financial resources, energy spending, logistical arrangements and regulatory powers. The structural preconditions of virtual worlds and networks are at the same time more complex and more demanding, and more relaxed. This fact derives from the productive connection and intersection between energy and information fluxes and from the corresponding autonomy (including the self-regulatory capabilities) produced through them (Maldonado, 1993).

5. Disembodied Actors?

We speak of disembodied actors (personae fictae including the individual and the complex organization) in relative terms, but in any case, the transformation is relevant enough to be emphasized and to generate distinct themes for social analysis. We now consider the position of such subjects in relation to the socializing impact of a networked and networking life. We do not yet know well how, but surely the fact itself of being basically formed through virtual processes and through networks will be a very different social thing than being socialized in face to face - or complex but localized - interactions. Assuming that difference as a given, we explore the capacity building component. What sort of persons, or of personae fictae, will result from such processes? Will they be able to grow up in autonomy, self-reliance, self-realization, or will they be slaves to the net, addicted or dependent on compulsory virtual worlds and on their unfathomable vocabularies of motives? Socially speaking, the dilemma has to be seen as a continuum going from serfdom (as well represented by addiction and drug-dependency) to liberty. The paideia of capacity building can only be an unfinished business, going from one body to another, from one identity to another, from one generation to another, and so on. Each individuation process is inherently ambiguous, perfection in identity may well be a crime, but progress and evolution are always possible (Sen, 1999; 2005). Capacity is the process of becoming a better person (in a normative sense, where the normativity must be recognized and, as far as possible posited by the subject itself), a person able to conjugate individuality and social bonds (Nussbaum, 2001; 2003). In these neo-Aristotelian terms, the question is now elaborated by Nussbaum and Sen concerning real persons or social subjects, without remarking on the peculiarities of the capacity building process for networked, and thus, disembodied persons.

6. Paideia and Capacities

The classical and humanistic idea of paideia, therefore, the idea of capacity or capability is a sophisticated derivation, always connecting body and identity. Paideia is to be understood as self-formation through social interactions. The concrete person, socially situated, is the actor who can strive for emancipation through paideia. The body must be healthy, and the mind alert. The paideia makes it possible to maintain a certain distance between the concrete person (as a unity of soul or mind and body) and the persona ficta, that will be constructed through social institution-building processes. In sociology, we speak of role distance to mean a certain degree of reflexivity in the contact between the person as a body-mind unity and the different social roles. The paideia itself is a weapon to avoid that the unity body-mind be totally submerged and annichilated in the social space. This is the critical point in the classical model, elaborated through the experience of the polis, of the res publica, but also through the turmoils of authoritarian regimes, as in Stoic ethics. A Stoic component has historically been added to the Aristotelian basis, positing the subject in a domain-free sphere, outside of excessively compromising power relationships. The distance, also the capacity to hold the distance from the socially constructed self and the true self, in other words: the necessity of distinguishing between self and society, is a relevant component, later on, in the thinking of Montaigne, of the French moralists and in the Scottish tradition till Smith and Mill. But to pursue such historical streams is not our concern here.

The classical subject is a body with a mind, or better the reverse. The idea of social actor – as in the sociology of M. Weber – holds firm to this tenet. The social actors that are socially constructed, such as complex organizations, are social analogues of the empirical subject. They too, as we have seen at the beginning, receive a body and a mind. Such unity is the precondition of identity. And identity, socially speaking, is all. Identity allows us to construct a world of discrete unities and to derive a Newtonian universe, moved by discrete unities mechanically connected, as is the case in the classical and neoclassical economic discourse. Sociology has always been uneasy with that, emphasizing rather the structural and emergent effects, latent functions, and collectives which are more than the aggregate of social atoms. How collective action is at all possible is the central theme of each sociological theory; that is to say, how is it possible for social atoms (unities of body and mind) to cooperate in producing social order. Collective action itself is not possible, the sociologist says, without a form of paideia. With individual emancipation goes the emancipation of freedom potentials in the collective. The body offers the needs, the mind offers the ideas: their coupling drives the entire process of emancipation and liberation. The process ideally goes through the continuum: body-mind-personal identity-social paideia, collective action-results in terms of freedom and autonomy both for individual actors and for the social collective. The paideia ripens in a general process of capacity building, expanding liberties and promoting potentialities. The social world ca not be perfect, but it can be perfected.

This central idea has been severely shattered by historical experience, but – postmodern foibles notwithstanding – the idea (in the humanistic sense of Panofsky) holds, as we can see in the thinking of Habermas, Rorty, Searle, or more controversely in Derrida or Nancy. The idea holds, insofar as there is no available rational substitute. But the idea can hold, only if we can go on carving our problems out of her. So now let's see what capacity will mean – in reference to paideia processes – for disembodied persons, or minds circulating virtually in networks, where the very body of actors becomes virtualised, producing, at the end, disembodied actors (Crossley, 2001; Turner, 1999).

7. Capabilities as Positive Freedom in Virtual Worlds

Capacity for the actor means autonomy or freedom. Capacity building processes enhance and enlarge the space and resources for exploring and implementing potentials. Two factors are relevant: a. the dotations, i. e. the bulk of socially meaningful resources for acting , and b. the entitlements, i .e. the socially practicable and legitimate claims (rights, but also needs that can strive for recognition). The mix of dotations and claims produces a space for acting on behalf of self-realisation, or functionings. The history of functionings for an actor is his/her history of freedom or serfdom. A collectivity or society of actors, with *satisfactory* functionings (i. e. functionings meeting normative standards, that in a democracy are normally set at the constitutional level as human or fundamental rights) build up a community of freedom and autonomy. In such a context, the process of capacity building will be recursively and reflexively fed, and thus, stabilized. Such an achievement is not possible without satisfying the basic requirements of subsistence (nourishment, health, housing and so on). The body of the actor is in a sense the protagonist at that level of capacity building through need satisfaction. Later on, at more advanced stages of the process of capacitation, an abler body will support the ripening of abilities, skills, virtues, civic manners and social bonds. Surely, the body will recede and the mind will take command. Thus, the paradigma shows a fidelity to the original classical statements on the body-mind relationship. But we must never forget the relevance of the satisfaction of basic needs – i.e. of an able body – for the ripeness of the mind. This aspect is understood well by thinkers such as G. Bateson (1980), who understand the mind as bodily grounded, and at the same time, transcendental.

In the process of capacitation, the body assumes an increasingly virtual form, as a component of the virtualities wherein the mind navigates: metaphorical, symbolic, communicative seas of sense, where the reflexivity is at home. When the virtual worlds become prevailing, as in the second or late modernity, the body itself will be more and more elaborated as a virtual hologrammatic object – as it is in many avantgarde art experiences. The meaning of the body for the actor becomes the body as object, resource and medium for a cultural transformation of the self. Yet, in its virtual form, the body impinges on the process of capabilities formation and elaboration. The chief aspect of this relationship between mind and the virtualised body is the body as the messenger between mind and nature, where nature means and implies our human nature (as modified both by nature and cultural evolution), our identity as nostalgia of concrete space and time coordinates, the outside nature as ecosystem, and Gaja as an ecology of natural and social games. It is as if the body, on the grounds of satisfied basic needs and of virtualising experiences, strives to communicate with the great chain of beings, and to find a place and peace -if at all possible- there.

Capabilities are liberties in the social world, and are peacemaking forces with nature through the cultural elaboration of the body. The sense of achieving more capacity includes coming of age, ripening, assuming responsibilities for all the virtual bodies elaborated by the society of minds (as in fact is going on in networking), and through that process, the writing of a new covenant with nature – as the big body – as anticipated by so many different thinkers, such as E. Morin, E. Wilson, H. Jonas, B. Latour (Latour speaks very pertinently of 'parliaments of nature', 1995; 1999). Only virtualisied bodies will be able to do that, when the body itself is a communicating organ and a cultural product for the mind. Capabilities will thus not be powers to control our internal and external natures, but ways to give more chances to reciprocal freedom (Krippendorff, 1999). Through such conumdrums, we will go back to nature, and in that measure, the mind will receive back its body as a medium both in society and in nature.

Persons as disembodied actors in virtual worlds, surely, but as a transition to a re-naturalized body and to more and more reflexivity in the mind-body relationship. Capabilities are here for that reason, as both Aristoteles and Sen suggest.

8. At the Crossroad

Disembodiment does not mean the disappearance of the body for either the conscious subject, or in the public scene. On the contrary. it is true that in virtual space, interactions and communication forms are possible and practised which do not imply the presence or symbolic reference to the very bodies of participants insofar as the stream of information, symbols, data, and images are set free to fly over and float through, and to agitate in a self-mirroring space without the delimitations typical of the 'real' world till to the topic moment of downloading. Here, the body re-appears more as a constraint than as a resource, as a slowing-down factor, as an impediment to be overcome. However, in virtual space, the transfigured, pixel- and bit supported body assumes a double functioning: *as an object of virtual manipulation* (any innovative sort of mutation), as a medium at the disposal of interactive experimentation, as an object between others, perhaps more communicative than others, or better, as a medium very apt and specifically adapted to communication and yet capable of multiple sense-making; and *as a plural being.* The first role is well known through videoart and the manifold world of video-mediated religious experiences (especially the masochistic 'rituals de passage' which acquire significance through their being multimedially recorded); the second one is reflected in virtual space- all the adventures and catastrophes of the plural self of the contemporary subject multiplied as in a mirror game. In real life, it can present itself – on the negative side - as a divided, spoiled self, while on the side of possible enactment, it is seen as a multiplied, or ripening and unfolding self. The first alternative is the most frequent in mass societies, following the analysis of Adorno, Bell, Sennett, or Lifton. The splitting of the self surely implies a suffering body, the body as a receptacle of malaise and unhappiness, which become here – through scientification and technical manipulation - clinical matters. *Das Unbehagen,* thematised by Freud, is dissipated today as a paradoxical quest for the quality of life in an affluent mass society. The multimirrored body becomes co-related to the multiple self through suffering or malaise; insofar as it begins a trip (a very topical word for virtual realities) through *terra incognita.* Such terrain is overtly ambivalent: virtualisation brings with itself the chance to escape from distress and unhappiness and to set up a multistaged agony. Or , on the contrary, to bridge the chasm between mediatisation and mediation (in its old philosophical meaning), spiralling up to ripeness. Let us not forget that, from another point of view, the body, through the economics of self-care (from diets to body building, from ayurvedic care to extreme sports) acquires grandeur, social representativeness, and excellence like animated self-advertising ware in the expanding universe of mercification. We can get happiness through wares, if the bodies are virtualised and meditatised in the same act. This implies the overall dominance of the form of ware in social worlds. This point is very important in understanding the deep transformation of our societies, the voices of alarm and protest, of dissent and conformism, the demand for surety, and the anxious quest for risks and assurances.

Taking this into consideration, we have to reflect carefully on the fact that virtual worlds are also a powerful medium for capacity building, because in this role they may find their appropriate historical or evolutionary legitimation. The plural self is developing through variety, shifting between the real (in the common sense) world and virtual space and times. The variety of time-space relationships and the very suspension of coordinates stimulate the intelligence of the actor, the trial and error, the muddling through, the learning process. A plural self, partially or momentarily disembodied thanks to diffused virtual experiences, is an experimenting one, that is unstable, but process-oriented. The point to find on that curve is the moment of reflexivity, this last glitter of the illuministic promise of happiness through ratio and practical reason. A partially disembodied self can be more flexible, adaptive and also opportunistic, but also more conscious of its limits, bindings, and intertwinings. In virtual worlds, there are disseminated opportunities to ripen as a subject, commons to be put to work, also rights or legitimate claims to be affirmed and powers to be contested. The network is here for implicit distributive justice. It can be used (it is only a possibility, but a very radical one) to build up capabilities, exercise positive freedom, expose illegitimate powers, and inappropriate and coercive tabus. The real point is how the plural self in connection (because the virtual body is a multimedium) with a pluralized body: partially real and empirical, partially virtual, mental, intentional, and iconic, can succeed in its relentless pursuit of happiness. That same pursuit is thought fundamentally as a process implying capabilities, better: capacity building.

The question to pose is: the very ideas of self, capability and positive liberty (that are also and substantively regulative ideas) were thought out for a consistent, self-consistent self, univocally determined in biography and preference, and strictly bound to its unique body. So long as we add liberties and positive freedom to the social aspect of the subject (a subject of rights first of all, and thus as *persona ficta*), we expand the sphere around the self and its embodied identity. Historically, all the single spheres expand and encroach with a growing number of other spheres and the incommensurable spheres of others through liberty, welfare, and material happiness. Freedom becomes the space free from – or built from - the interference of individual spheres. Furthermore, in that developmental and quantitative expansion, the subject is assumed to be coherent, in a sense monadic, like a steadfast anchorage at the very centre of social relationships. Institutions can care for the variety of selves and also for the proliferation of body forms, but identity is still posited normatively as unique, to impute responsibility, consequences and projects as well as to measure advancement into the space of capabilities. Till now, we can also take into account this initial proliferation of the self, constructing multidimensional analysis instruments, evaluation criteria, multipurpose curricula, and multilateral biographies. In any case, we (in the West, I mean) normatively hold to the possibility of a reduction to one of the multiplied bodies and selves, or to discover the true self (and the true body?). The pursuit of happiness is indeed often described as a quest for authenticity, from Romanticism on. The ripening process of individuals described as a search for authenticity means that somewhere, what is true or original in the self and for the body must exist: something authentic to discover among the different and the increasing number of available and disparate alternatives.

Yet, we are engaging in a much harsher evolution through the disembodiment of self in virtual worlds. To cope with the realities of virtualised subjects, we have to develop conceptual tools and institutional resources adequate in sustaining and 'rendering' more reflexive the incommensurable plurality we encounter in the potentially desocialised virtual space. The paideia of new self-body configurations is yet to be formulated, but in the field, the experiments are going forward, overcoming the same apories of a reductive postmodernist outlook. This has offered an initial, but distorting, language to cope with increasing variety and virtualisation. We have to go back to the basics. There is something classical (that means: norms awaking) in virtual worlds and in the disembodiment processes, an aura we know well from the acme of art and thought, from the *revoltes logiques* of awe-inspiring modernity. All the paideia (here: truly to be intented as an Oriental-Occidental *Diwan*, as Sen has forcefully argued) imply the unfolding of potentialities. The classical supremacy of mind over body can be translated into a new alliance between virtualised bodies and pluralized minds, an alliance between abstract analytic semantics and image proliferation. Both the growing abstraction required to cope and to comply with complex worlds and the flickering of ubiquitous multisensorial streamings evoke, provoke and invoke potentials, their unfolding and valorisation: any thinkable paideia, if any, is here for that. The differentiation through virtualisation processes can construct a new idea of subject, of a multi-level persona ficta, one much more universalised and universalistic than the usual social actor, that is so emptied out of levels, dimensions, attributes and potentials. Because there is only a slim chance, we need to be careful – of both minds and bodies. Therefore, we are left with many yet unanswered questions: what sort of togetherness and social bonds, which arenas and agoras do we need to communicate variable, dissipative identities, while we are experiencing the dissolution of monads in the plurality and multiplicity of self/body, real and virtual, connections, where the only constants are the very couplings of selves and bodies: everchanging, stochastic and precarious? How are partially or cyclically disembodied entities searching for selves and bodies with their yet unnamed configurations - oversocialized actors within oversocialized and, therefore, partially virtualised bodies - struggling for life and meaning -, and thus, donating life to the expanding galassia of virtual worlds?

References

AA. VV., (2004), Umano – post-umano. Roma: Editori Riuniti.

Adorno, Th. W., (1967), Negative Dialektik, Frankfurt: Suhrkamp.

Barabasi, A-L. (2004) Link. Torino: Einaudi.

Bateson, G. (1980) Mind and Nature. London: Fontana/Collins.

Bazzicalupo L. – Esposito R., eds, (2003), Politica della vita, Roma-Bari: Laterza.

Bell, D., (1976), The cultural contradictions of capitalism, New York: Basic books.

Beck, U., (2003), La società cosmopolita, Bologna: Il Mulino.

—. (2002) Macht und Gegenmacht im globalen Zeitalter. Frankfurt: Suhrkamp.

Boltansky, L. and Chiapello E. (1999) Le nouvel esprit du capitalisme. Paris: Gallimard.

Boltanski, L. and Thévenot, L. (1991) De la justification. Paris: Gallimard.

Burke, K. (1962) A grammar of motives. A rethoric of motives. New York: Meridian Books.

Castells, M. (1996) The rise of the network society. Oxford: Blackwell.

Crossley, N. (2001) The social body. London: Sage Publications.

Crossley N. and Roberts, J. M, eds, (2004), After Habermas, London: Blackwell.

Descombes V., (1996), Les institutions du sense, Paris: Les Editions de Minuit.

Donolo, C. (1997) L'intelligenza delle istituzioni (The intelligence of institutions). Milano: Feltrinelli.

—. et al. (2005), rete, parolechiave, 34, Roma: Carocci.

—. (2005), 'Networks as commons', parolechiave, 34, (Carocci editore, Roma)

—. (2006) Going on lightly constructed bridges. (to be published) Milano: Feltrinelli.

—. (ed.) (2006) Il futuro delle politiche pubbliche (The future of public policies). Milano: Bruno Mondadori.

—. (2006), 'Nuovi soggetti per nuovi diritti', in Fondazione Basso, Dichiarazione universale dei diritti dell'uomo, Roma: esi.

Friedmann, J., (1994), Cultural identity and global process, London: Sage.

Krippendorff, E. (1999), Die Kunst, nicht regiert zu sein. Frankfurt: Suhrkamp.

Jullien F., (2005), Nourrir sa vie à l'écart du bonheur, Paris: Editions du Seuil.

Latour, B. (1995) Non siamo mai stati moderni. Milano: Eleuthera.

—. (1999) Politiques de la Nature. Paris: Editions de la Découverte.

Lifton R. J., (1993), The protean self: human resilience in an age of fragmentation, New York: Basic Books.

Maldonado T. (1993), Reale e virtuale, Milano: Feltrinelli

Nussbaum, M.C. (2001) Diventare persone. Bologna: Il Mulino.

—. (2003) Capacità personale e democrazia sociale. Reggio Emilia: diabasis.

Ostrom, E. (1990) Governing the Commons. New York: Cambridge University Press.

—. (2004) Understanding Institutional Diversity. Downloadable at www.indiana.edu.

Perulli, P. (2001) La città delle reti. Torino: Bollati-Boringhieri

Rodotà, S., (2006), La vita e le regole, Milano: Feltrinelli.

Schelling, Th. (1978) Micromotives and Macrobehavior, Cambridge, Mass.: Harvard University Press.

—. (1984) Choice and Consequence, Cambridge, Mass.: Harvard University Press.

Sen, A. (1992) Commodities and Capabilities. Amsterdam: North-Holland.

—. (1999), Development as freedom, New York: Alfred A. Knopf.

—. (2006), Identità e violenza, , Roma-Bari: Laterza.

Sennett R. (1990) The Conscience of the Eye. New York: Alfred A. Knopf.

—. (2006), The culture of the new capitalism, New Haven:Yale University Press.

Taylor, M. (2003), The Moment of Complexity: emerging network culture, Chicago: The University of Chicago Press.

Teubner, M. (1993), 'Thew State of Private Networks', in Brigham Young University Law Review, 2.

—. (2004) Societal Constitutionalism. At. www.wisc.edu/wage/papers/teubner.

Turner, B.S. (1999) The Body & Society. London: Sage Publications.
White, H.C. (2002) Markets from Networks. Princeton: Princeton University Press.

MIGRATION, POLITICAL CULTURES AND URBAN CONFLICTS IN EUROPE

UMBERTO MELOTTI

Abstract: Immigration has introduced serious conflicts into European urban areas. The nature of these conflicts varies according to many factors: the countries of origin of immigrants, their cultural and religious heritage, the social and economic problems in the hosting countries. But it is also necessary to take into account the political culture of these countries, i.e. the way in which State, people and nation are regarded and the relationships between ethnicity, nationality and citizenship are conceived. It is not by chance that for long France has adopted a policy of integration through cultural assimilation, the United Kingdom has practised a form of uneven pluralism emphasizing the role of ethnic communities and Germany has preferred a rotatory presence of 'guest workers'. Yet, all of these policies have failed owing to the important changes that have occurred both in the global context and in immigration itself. This paper analyzes what has happened by taking into account the urban riots in England since the late '50s, the xenophobic attacks on immigrants in Germany after its reunification and the recurrent explosions in the French *banlieues* since the late '70s until the resounding events of November 2005. The last part concerns the case of Italy, which is still virtually unknown abroad.

Keywords: Conflicts, Immigration, Europe, Britain, France, Germany, Italy.

1. Introduction

November 2005. Paris burns owing to the 'intifada of the *banlieues*': a rising, a revolt, a rebellion, an insurrection, or even a new 'vertical invasion by the barbarians', as it has variously been called by the mass media. Protagonists of this 'collective action' were, once more, but on an unprecedented scale, young immigrants and children or grandchildren of immigrants from the former African French colonies. Most of them were 'Maghrebians' — as the French define the people of Northwest Africa and their descendants — or, rather, *beurs*, as they call themselves in their slang (the *verlan*, the *langue à l'envers*), by reversing and mangling the word *arabes*: almost a symbol of their divided and distorted identity. But among the rioters there were also many young people from the subSaharian countries, the *blacks* (as they call themselves): a fact quite new, at least on account of their number, which was not emphasized by the first observers.

The revolt, which lasted for almost one month, broke out after an accident that provoked strong reactions among the young people of the *banlieues*. On October 27 2005, in Clichy-sous-Bois, a municipality of the department Seine - Saint-Denis, in the Paris metropolitan area, three young men of migratory origin, who believed they were being pursued by the police, tried to hide in the cabin of a transformer, where two of them were electrocuted. The news spread quickly, giving rise to extensive rioting throughout the country. The consequences were serious: thousands of cars were burnt or destroyed, there was pillaging, arson, vandalistic acts of various kinds as well as many wounded and injured and even one death. This induced the government to impose a curfew, the first in France since the Algerian war, which remained in force until the new year.

Politicians and scholars interpreted these events in various ways. The then French President, Jacques Chirac, attributed it to the young immigrants' 'identity crisis'. His successor Nicolas Sarkozy, then Interior minister, blamed the bad attitude of this 'scum' (*racaille*), a term that exacerbated the spirits and inflamed the situation. The philosopher Alain Finkielkraut, an ardent supporter of the secular 'Republican' values, spoke of an ethnic and religious revolt, due to the bad influence of Islamic preaching ('They are not poor, they are Muslims'). In contrast, Tariq Ramadan, a prominent European Muslim scholar, defined it as a consequence of the discrimination suffered by all young people of the banlieues, regardless of their race, religion and culture. The sociologist Alain Touraine underlined the lack of social integration of immigrants, confirming his previous analyses of some similar events. The anthropologist Marc Augé attributed this 'protest without ideological contents and direct relationships with political Islam' to the competition between youth gangs of the suburban areas, which, according to him, with spectacular acts of violence, tried to gain space on the television screen, the new multiplied, fragmented and scattered centre of everyday life. Other scholars put forward interpretations in terms of 'inassimilability of new immigrants', 'youth problems', 'new poverty', 'urban question', etc.

However, none of these analyses is entirely satisfactory, since it is necessary to take into account all of these factors and many others, including those brought to light few months later (March 2006) by the movement of the students (most of them *français de souche* and not coming from the *banlieues*) against the law on work placements, blamed for favouring precarious conditions.

Anyhow, in the French events *de te fabula narratur*. Even Italian politicians, who are not usually very quick on the uptake, understood it quite well. Romano Prodi, at that time recently re-nominated for the premiership by the centre-left coalition, warned against the risk of similar outbursts in Italy, while Roberto Calderoli, a leader of the Northern League, then minister in the centre-right government, took this opportunity to rail once more against the dangerous consequences of the Muslim 'invasion'.

Indeed, the events of November 2005 do not concern only Paris, or even France, which, however, had already experienced similar outbursts, though on a minor scale: in 1979, 1981 and 1990 in the *banlieues* of Lyons, in 1991 in the area of Paris, and later in many other town of any size, including Marseille, Nice, Strasbourg and Bordeaux.

In the United Kingdom rioting by 'ethnic minorities' (as the 'black' and Asian immigrants and their descendants are usually defined) has exploded many times in the suburbs of London (since the late '50s) and, later, in the inner cities of Manchester, Liverpool and Birmingham. Many other towns have experienced similar outbursts: among them, Bradford, Bristol, Brixton, Oldham, Leeds and Burnley.

Germany, too, has experienced very serious clashes between nationals and non-nationals, but mostly of quite a different nature. In fact, they were due much more to the xenophobia of the former rather than to the rage of the latter. Especially after the reunification of the country (October 3, 1990), beginning from the towns of the former German Democratic Republic, the attacks on immigrants spread all over the country and many people perished when their hostels were burnt down. People were dismayed by these dreadful crimes and, after one of the most terrible, the president of the Federal German Republic himself attended the funeral of the victims not only to express the official mourning of the country, but also to underline the need to change policy for both immigration and citizenship rights. This new course, however, began only some years later, and not without serious difficulties.

After September 11 2001 the situation worsened throughout Europe, owing to the increasing menace of Islamic terrorism, whose devastating attacks struck two important European capitals: Madrid, on March 11 2004, and London, on July 7 2005. Before that, on July 27 and August 17 1995, Paris had also been hit by Muslim terrorists, belonging to the Islamic Salvation Front, an organization that had already killed many thousands of people in Algeria.

Rome, also as the main seat of the Catholic Church, is particularly exposed to this kind of risk (as have recalled the very serious threats uttered by some Muslim exponents after Benedict XVI's Ratisbon lecture in 2006). After the attempt on John Paul II's life in St. Peter's Square by a Turkish assassin (1981) and the attack on the El Al desk in Fiumicino Airport by a Palestinian commando (1985), other attempts had been foiled only by chance. The same occurred in other Italian towns, such as Milan, Naples and Bologna. Some Italians, however, were killed abroad, where they worked or where they were spending their holidays. Others were very lucky to survive awful slaughter or to be spared from lynching, as occurred in Bengasi, on February 16 2006, during the assault on the Italian consulate by a fanatic mob, infuriated by the 'Islamic cartoons' flaunted on his T-shirt by a member of the Italian government.

Other European countries also experienced some bloody deeds more or less directly connected with immigration. In the Netherlands, for instance, on December 2 2004 the director Theo Van Gogh was killed by a fanatic Muslim of Moroccan origin, who intended to avenge the 'offence to Islam' that, according to him and other extremists, had been caused by his film *Submission*, which denounced the condition of Muslim women.

Such extremists are also present in other European countries and Italy is no way excluded. Suffice it to say that a sociologist of Egyptian origin, Magdi Allam, one of the vice-editors of the 'Corriere della Sera', the most important Italian newspaper, is now obliged to move with an armed escort on account of his outspoken warnings of the close relationships between the imans of many Italian mosques and some terrorist organizations.

2. Immigration and Urban Conflicts: Some Exemplary Cases

Important urban conflicts closely connected with international migration have emerged since the '80s in the traditional countries of immigration on both sides of the Atlantic.

An exemplary case is that of Los Angeles, the greatest American town on the western coast, where an increasing number of immigrants from Asian and Latin American countries join the native 'blacks', who still live, in many cases, in grim conditions of social exclusion.

In May 1992 the South Central - Watts area of Los Angeles was hit by a series of terrible riots, resulting in 650 fires, pillaging and destruction, which left over 2,300 wounded and 58 dead. These riots, like many others of the same kind, broke out owing to an event of limited importance, though very unpleasant: the acquittal of four police officers who had beaten up a young black motorist with a criminal record who had failed to stop when flagged down. The exceptional violence of the revolt made this town the symbol of the new urban conflicts that recall the scenarios of *Blade Runner*, the cult movie set in the very city of Los Angeles, at some time in a near future: an alienated and anomic Babelic metropolis, where separated and conflicting groups live as foreigners to one another and even to themselves, segregated in different parts of its territory, in conditions of urban degradation, marginality and social exclusion.

Similar events, however, had already taken place in other U.S. towns, such as Detroit, Miami, New York and Washington D.C. Here violence had exploded in many forms in the huge ghettos that aggravate the sense of deprivation and captivity in areas that appear to be assigned to their inhabitants by poverty (a quarter of their population was officially classified as poor at the beginnings of the '90s) and ethnic origins (a result, for many of them, of more or less recent immigration): the same factors that had already hampered the integration of the previous generations and had favoured the proliferation of youth gangs engaged in continuous bloody conflicts (Body-Gendrot, 1993: 36-40).

The persistence of this situation was recently confirmed by the events that accompanied hurricane Katrina and the subsequent floods, which, between August and September 2005, devastated New Orleans and killed thousands of people. In this town of about 500,000 inhabitants (over 70% 'non-white', according to the official data), pillages of shops and supermarkets and clashes between gangs and the army and between gangs from various suburbs brought to light pre-existing conditions: a vast urban poverty (involving at least a quarter of the population), racial exclusion and ethnic segmentation, worsened by the effects of the pattern of life fed by the existing social system, its ideology and its scant social policy. The disaster, as Franco Ferrarotti remarked (2005), showed once more that the poor always pay twice: first, because many of them live on the threshold of a precarious survival; second, because they are the ones most exposed to the effects of calamities. But to speak only of 'poor people' in this case would be misleading. Almost all victims were black, creole or half-breds: 'Among them there were almost no white. Why? The newspapers say that the order to leave the town had arrived too late. But for most black people and all the 'unprivileged', as they are hypocritally defined in official language, that order would have never arrived in time, because they had no personal means of transport to save themselves'. On the other hand, the poor of New Orleans were not a class, in the European sense. 'They were, rather, a vast and various underproletariat, obliged to live by its wits. Therefore, their propensity to pillage, though not justifiable, was quite understandable. The same facts had occurred in all previous riots or revolts, especially in the ghettos of the greatest American towns, such as

New York, Chicago and Los Angeles. Some observers unaware of the real situation denounced that 'anarchy'. But I would rather define it as the grotesque and desacrating feast of the poor, who were emarginated and excluded from the American dream'.

A few days after the Los Angeles riots one of the most important French newspapers, 'Le Monde' (May 16, 1992), in an article signed by the mayor of Lyons, after recalling the riots which had occurred in France, Britain and South Africa (in Soweto, the large black ghetto in the City of Johannesburg), wondered why people were so blind as not to see the global urban crisis that was becoming 'the most important political issue' in many parts of the world.

The most prestigious English newspaper, 'The Times', did not wait for the Los Angeles revolt to raise the alarm. Already on July 27 1967, comparing the riots that had appeared in some British towns with those that had taken place in the United States, it wrote: 'Birmingham is no Detroit, but there are storm signals [...]. Race riots in North America raise an anxious question for some people living in Britain's immigrant areas: could it possibly happen here? Some people who know race relations think the answer is yes'.

It is difficult not to appreciate this early warning, in spite of its reductive formulation in racial terms (but in Anglo-Saxon countries the word 'races' is often improperly used to define the 'ethnic groups'). In fact, urban riots not so different from those that had taken place in the United States have already occurred in many European towns. Especially in France, Britain and Germany these conflicts have been quite numerous, even if in each of these countries they have assumed specific traits.

2.1 The Urban Conflicts in the United Kingdom

Since the early '80s in the United Kingdom inter-ethnic conflicts have not only been occasional events, limited to a few districts or towns. Actually, they have been an expression of the social situation of the country, as have shown the numerous riots that have repeatedly taken place in almost all English towns: those that had been most affected by the process of deindustrialization, such as Liverpool, Manchester and Birmingham, as well as London itself, especially some of its districts and suburbs.

One of the most serious riots took place in autumn 1985 in Birmingham, where 2 people were killed and 122 were wounded. Also in this case an event of scant importance, a fine inflicted on a coloured driver, was sufficient to cause many cases of arson and pillaging. The same surprising disproportion between the precipitant event and its consequences characterised the riots in Brixton, Tottenham and Toxteth.

According to the commission that investigated the 1981 disorders in Brixton ('the worst in the United Kingdom' in the last century), the revolt broke out after police action that was perceived as racial harassment. But its underlying cause was the objective 'racial disadvantage: a fact of British life' that was necessary to eradicate, in order to prevent its transformation into an endemic plague that could undermine the social organisation of the country (Scarman, 1981).

Analyses of this kind were substantially repeated in occasion of subsequent urban riots, in spite of the important legislative and administrative measures that had been taken in the meanwhile to combat racial discrimination and to promote equal opportunities for all ethnic groups. It is not by chance that at the beginnings of the '90s a keen observer of the British social life affirmed that 'Britain could become a successful multi-racial and multi-ethnic society', but it could also evolve in the opposite sense, with devastating effects not only on the condition of immigrants, but on civil life itself (O' Donnell, 1991: 37).

Since then not all doubts have been removed and other serious urban conflicts have taken place in Manchester, Burnley, Oldham, Leeds and especially Bradford, a town with an Asian population of about 100.000 people. Nevertheless, today the situation could justify a certain optimism, if a new risk had not appeared on the scene: Islamic terrorism. In 2005, on July 7, a group of suicide bombers of Asian origin born in England caused havoc in the British capital and two weeks later, on July 21, another similar group tried to do it again. Therefore, it is not at all reassuring that, in a survey carried out a few days after these attacks, 88% of the Muslims living in Britain, including those who were born there, affirmed that they do not feel really British.

Afterwards the arrest of 24 young British citizen of Pakistani origin and Muslim religion who were on the point of carrying out suicide attacks against six airliners in departure from Britain to the United States (August 2006) and two attempts foiled in London and another one partially failed in Glasgow (June 2007) have alimented further doubts on the British pluricultural model, which was blamed for favouring the formation of real 'Londonistans' in a country were there are above 1,800,000 Muslims and 1,500 mosques. Anyway, Britain reared those 'robots of death' who were ready to sacrifice their life in order to strike the West they hated.

2.2 The Urban Conflicts in France

In France, too, ethnic conflicts have been quite numerous since the late '70s and have repeatedly occurred in all the main towns.

The first severe clashes of this kind appeared in the 'hot summer' of 1981, when rioting exploded in Vénissieux, Villeurbanne and Vaulx-enVelin, in the urban area of Lyons. These municipalities were all characterised by high percentages of marginal populations, mostly of foreign origin, living in conditions of serious economic and social deprivation, unemployment or underemployment, bad housing and low levels of school education. The fighting began when in Les Minguettes, a *cité* of Vénissieux, the police intervened against a group of young people (mostly immigrants or children of immigrants) who used to hold 'rodeos', as they call these 'games' where stolen cars are driven at break neck speed in inhabited areas and are then smashed against walls or other cars or are burnt. These riots, which received much attention in the mass media, forced the French to discover the problems of the *banlieues*, the 'relegation', 'exile' or 'segregation' areas of their towns (Delarue, 1991, Dubet, 1992, Gallissot and Moulin, 1995, Donzelot, 2006), inhabited by a high percentage of young people of foreign origin, without skills or a profession, and often addicted to drugs or drinking, who were considered, for good reasons or prejudice, responsible for the sale of drugs, theft, burglary, rape, vandalism and many other crimes and offences.

The police, who keep a close eye on them, when they have grounds to do so do not hesitate to intervene with a heavy hand, especially against the young people of African origin. When it happens, the latter and their supporters, in good or bad faith, cry out against the 'racist repression'. After 1981 the protests of this kind multiplied and eventually gave birth, in December 1983, to a 'March for Equality' that left Marseille with only a few hundred participants, but arrived in Paris with over 100,000 people. Afterwards, building on its success, its organizers formed the two main French 'anti-racist' associations, Sos-Racisme and France-Plus, which worked efficiently to improve the conditions of immigrants. Nevertheless, in the following decade a new phase of ethnic and social conflicts began. Large-scale disorders broke out in 1990 in Vaulx-en-Velin, in the *banlieue* of Lyons, and in 1991 in Sartrouville, in the *banlieue* of Paris, and in Mantes-la-Jolie, a small town about 60 km from the capital. Some years later, after a period of only apparent tranquillity, which however concealed a worrying criminal drift, there was the great explosion of November 2005 and fighting broke out all over the country.

During the 1990 riots in Vaulx-en-Velin, I was in France carrying out a part of an exacting research project on immigration in Europe (Melotti,

1992a) and I went there immediately. Two years later I returned to Vaulx-en-Velin, when carrying out a comparative research on immigration in Milan and Lyons. Thus I could investigate those events at first hand (Melotti, 1992b).

The riots began on October 6 1990, on Saturday, after the death of a young man, son of Spanish immigrants, who was involved in a controversial road accident: according to the police, it was due to the irresponsible behaviour of two young motor-cyclists, who had run off to avoid a routine control; according to the young rioters, it was caused by to an unjustified intervention by the police. The protests for that death soon degenerated into heavy clashes, so much so that the situation was defined in terms of urban guerrilla warfare, revolt, rising, insurrection. Certainly, there was an explosion of rage accumulated over time by the young people of foreign origin living in that area, which was not very pleasant, though less severe than many others of the same kind.

Vaulx-en-Velin, the fourth largest municipality of the department of Lyons, then had over 45,000 inhabitants, after its population had doubled owing to massive immigration of the previous decades. Its population was quite young (50% of its inhabitants were under 30 years of age) and was mainly formed of immigrants or children of immigrants, mostly from the Maghreb countries, in great part (75%) living in the very quarter where the riots broke out. In its territory there were more than 1,000 private businesses, which employed a total of 13.000 workers, and 22 schools, but none of them was a *lycée*. More generally, its demographic and economic growth had not been accompanied by an adequate development of infrastructures and services. Moreover, there was considerable unemployment, due to the long-standing process of deindustrialization in the region, and young people ran the risk of remaining jobless when they left schools. The municipality was run by the left, but very few people appreciated its work. Among the previous events that could explain the exacerbation of the young, we must recall several incidents with the police, with eleven victims, ten with foreign names.

To this 'weekend of fear' all the French press devoted considerable space. At first it emphasized the events with banner headlines on the first page and more or less dramatic articles on the 'hell of the *banlieues*' and the 'drifting mine' of immigrants; then it also hosted some articles by sociologists and other scholars. The prevailing interpretation was in terms of 'untimely reaction of disadvantaged communities': the same that had already been formulated for the clashes of the previous decade. But there were also more complex analyses.

Some authors, by recalling the importance that assimilation assumed in the French model of integration, attributed the situation to the alleged 'inassimilability' of the new immigrants, mostly coming from Maghreb and subSaharian African areas. To quote Max Clos ('Le Figaro', October 17 1990): 'The key problem is the presence in our territory of inassimilable communities. The violence in the *banlieues* is not due to extraterrestrial beings. Both the victims and the police know very well who are its responsible. According to the different geographical areas, they are the blacks or the Maghrebians, as the pictures transmitted by all televisions clearly show. Some people try to convince us that this is the (just?) revolt of disadvantaged communities. But such an explanation is quite unsatisfactory. In France there are other immigrants (such as the Portuguese and the Asians) who experience the same difficulties, but do not raise problems either for public security or in everyday life. Why is there this difference? Because these communities desire integration and try to achieve it. The Maghrebians, on the contrary, live in France, but do not recognize its values and refuse its norms'.

This explanation, not only alarming, but 'politically incorrect', was harshly criticised by many people. In contrast to it, two well-known specialists of migratory problems, Antonio Perotti and France Thépaut (1990), proposed an interpretation based on the 'urban question'. By building upon the conclusions of an important official report on the *banlieues*, they affirmed that the conflicts in these areas were not due to the alleged cultural 'inassimilability' of the new immigrants, but, rather, to their scant social integration. According to them, it was the urban structure itself which impeded the insertion of their inhabitants into a normal life and aggravated social tensions. Therefore, to solve the problem, it would be necessary to dismantle these dormitory towns or, at least, to introduce there productive activities and living centres to transform them into 'quarters like the others'.

This interpretation, too, has, however, its weak points. Actually, Vaulx-en-Velin was a good example of the urban renewal, carried out there since the mid '80s. Exactly for this reason another scholar, Adil Jazouli, a sociologist of Moroccan origin who had founded an observatory expressly devoted to social life in the *banlieues*, advanced a different explanation. By recalling the results of his previous research on the gangs of young immigrants (mostly Maghrebians, subSaharians and Caribbeans) that had appeared in France in the '80s, he proposed an analysis in terms of 'new youth question': the riots in the *banlieues* were mainly due to the sufferings of the new multiracial young proletariat, lacking in citizenship rights and social relationships, which had formed in France owing to immigration ('Le Monde', October 16 1990). In his opinion, the policy for the integration of immigrants and their families that France had pursued with undeniable efforts since the middle '80s was quite wrong or, at least, insufficient. It was aimed at integrating the small groups that had promoted the movements in the '80s (an *élites* that had already become a sort of *beurgeoisie*, i.e. the 'bourgeoisie' of the *beurs*), neglecting the younger immigrants, who continued to live in conditions of poverty and social deprivation. According to him, the savage violence of the latest riots was caused by some of their groups, who intended to attract the attention of a public otherwise completely indifferent to their problems.

Yet, also this analysis is unsatisfactory, for it ignores the direct political aspects of the problem. These were underlined by a philosopher, Jean-Paul Dolle ('Liberation', October 15 1990), who wondered why some people could still be surprised by the riots in the *banlieues*: 'Whoever has set foot in a French *cité*, even if he does not know *when*, must know *where* and *how* these explosions may occur'. In these areas, according to him, the limit of resignation had already been exceeded and the young people lived in a situation of potential revolt. Inequality had become greater and greater, but nobody had done anything to contrast it. Moreover, the political system had become increasingly similar to that of the old Greek towns, where democracy was reserved to citizens and tolerance to the wealthy alien residents, while the poor foreigners were treated harshly. *De facto*, in France there had even been a return to census suffrage: poor immigrants, even if entitled to vote, tended not to exert this right. The consequence was an explosive development of social strata without real citizenship rights in urban zones more and more similar to ghettos.

A more complex interpretation was worked out by Alain Touraine, a sociologist well-known for his studies on social movements. The events of Vaulx-en-Velin, he wrote ('Liberation', October 15 1990), questioned the French model of integration through cultural assimilation, which had become obsolate. Ethnicity, a factor not foreseen by that model, had already burst into the French scene, but the only answer that it had received was social exclusion. The Vaulx-en-Velin riots, which had not been directly caused by poverty or isolation, but by an accident between immigrants and police, had a clear-cut ethnic character. Anyhow, ethnicity was one of the very few factors that could still produce important collective actions, after the collapse of the mobilising power of political parties and trade unions. The new ethnic conflicts, however, were quite different from the old ones (e.g., the attacks on Italian workers in southern France in the late nineteenth century or the *progroms* against the Jews before and after the First World War). They were due not only to the scant integration of ethnic minorities, but also to the change that had occurred in the social structure itself. From a society that was extremely stratified owing to the existence of various social classes, France had become a middle-class society, where however the excluded had considerably increased, for integration and exclusion had grown together.

Touraine also dwelt upon the change which had occurred in the identity of social actors: "We have learnt to define human beings for what they do, and not for what they are. But the new actors define themselves for what they are: women, young people, regions, nations, ethnic groups renamed communities. The decomposition of the social and political life leaves the field open to the globalised economy, on one hand, and, on the other, to the communities, defined by their peculiar traits: their traditions and their acceptance or refusal".

In his opinion it was not possible to give an answer to the new problems simply by trying to increase or decrease the rhythm of change. The result would have been a war of everyone against everyone else. On the contrary, it was necessary to recognize cultural diversity and to fight against social exclusion: two closely connected objectives. Therefore, he harshly criticised the policies that favoured the middle class alone, especially in education.

In a subsequent contribution, Touraine (1991) expressed these ideas even more clearly. In the old liberal France, he wrote, there was inequality, but not segregation in urban ghettos. Now, by developing under the influence of the American model, France was becoming a segregating society. The same process was taking place also in other European countries; yet, in France it was more dangerous, owing to the strong centralization of the country: 'Our centre creates barriers, such as that between Paris and its periphery: a symbolic distinction without equivalents in the world'. This worsens the contradiction between the lack of social integration and the policy of cultural assimilation officially pursued.

This diagnosis was substantially confirmed by the new wave of riots that exploded in France in 2005 and 2007. These conflicts, in fact, have made all too evident the limits of the French model of integration, which envisages the formation of a 'community of citizens' formally equal in their rights, without doing anything to face urban segregation and ethnic and social discrimination (Lagrange and Oberti, 2006). What is more, recent urban policies have even aggravated the situation (Donzelot, 2006).

Yet, the increasing capacity of exclusion of the new globalized society is a general trait that exacerbates conflicts almost everywhere: in the towns of the centre of the world system as well as in those at its periphery.

2.3 The Urban Conflicts in Germany

The situation of Germany is quite specific. There a great wave of urban conflicts exploded soon after its reunification, but, as mentioned above, generally they were clearly due more to the xenofobia of Germans than to the rage of immigrants.

This xenofobia was partly due to German political culture. In Germany, which was the last of the great western European countries to be constituted into a nation State, 'belonging' to the nation — far from being conceived in subjective civic and political terms, as in France, where Renan defined it as a sort of 'daily plebiscite' — has always been conceived in objective, ethnocultural terms, as a fact linked to blood and land (*Blut und Boden*) and to the putatively irreducible specificity of the German people (Melotti, 2006).

However, to understand these conflicts, it is useful to recall the situation of the late '80s and the early '90s, when the crisis of the Eastern European countries precipitated a flood of asylum seekers into West Germany. More than 1.5 million people – with a net total of around one million – arrived between 1989 and 1990, before the reunification, counting both the *Übersiedler* (the German refugees coming from Eastern Germany) and the *Aussiedler* (the descendants of the Germans who had settled many generations before in some eastern European countries). The absorption of the German Democratic Republic into the German Federal Republic in 1990 further complicated the situation by raising quite a number of new problems, due to internal migration from the new *Länder* to the old ones and to the tensions that emerged between the Germans of both the new and the old *Länder* and foreign immigrants. In subsequent years the arrival of *Aussiedler* did not cease (222,000 in 1991 and 230,000 in 1992) and those of other asylum-seekers kept on increasing (from 193,000 in 1990 to 256,000 in 1991 and 438,000 in 1992) until new restrictive measures were passed in 1993. In this context racism and xenophobia became a very serious threat throughout the country.

The attacks on immigrants began in towns of former German Democratic Republic (the first to experience them was Rostock, the main sea port of Eastern Germany). Yet, they soon spread also to Western Germany, where the old policy aiming at keeping immigrants in a precarious situation reinforced division, ill-feeling, prejudice and hatred.

Afterwards there were some signs of resipiscence about that policy, though even the most authoritative statements in favour of immigrants' integration revealed a rather reductive view of its nature and function. In fact, integration was not conceived as a spontaneous consequence of the development of normal social relationships between people of different origins, but, rather, as a process dictated from above, aiming, first of all, at ensuring the interest of the German citizen, eager to use immigrants' work without the costs and risks deriving from citizenship. It is, therefore, not surprising that in those years there was even a return to the old policy of 'guest workers' (*Gastarbeiter*), especially for the workers coming from Poland and other Eastern European countries.

One of the most important signs of change was the forerunning speech made by Richard von Weizsäcker (1993), the then president of the German Federal Republic, in the main mosque of Solingen, in front of the biers of five Turks killed in the most cruel racist attack of that period: 'Do the extremists claiming 'Germany for Germans' want to change our Constitution? For it does not say *Germans'* dignity is inviolable; it says *human beings'* dignity is inviolable [...]. As to the Turks [*living in Germany*], would it not be much better to begin to consider them German citizens of Turkish origin?'.

A great step forward in this direction was the new citizenship law, which was passed in 1999 by the new leftist majority, in spite of the strong resistance by the opposition. It was followed in 2004 by an even more disputed law on immigration. A first bill had been passed in 2002 by a narrow majority, but, before coming into force, the law was annulled by the Federal Court, which upheld the claim lodged by the opposition for a serious procedural mistake in its approval. Finally, after long discussions, it was substituted in 2004 by another bill, much less exacting, which was passed with the consent of the opposition.

However, even before these laws, some Germany towns had taken important social measures to face immigration. It is worth mentioning the case of Frankfort, a great town with a high percentage of immigrants (26%), where Daniel Cohn-Bendit, who was for a long time its councillor for multicultural affairs, instituted a pioneering board for intercultural mediation. Its aim was to prevent or reduce conflicts, both between German nationals and immigrants and between different groups of the latter.

Since then the social policies for immigrants have undergone an important development also at the federal level, due to the influence exerted by the EU institutions, which had begun to call for 'social integration of immigrants with respect for cultural diversity' (European Commission, 2000): an approach that was even proclaimed by the EU Constitutional Treaty, which was solemnly signed by all member States on October 29 2004, but did not come into force, because its ratification was later rejected by referendum in France and the Netherlands.

Yet, in spite of all the reforms mentioned above, in Germany the xenophobic attacks on immigrants have not completely ceased, even if the press now tends to speak of them as little as possible. This has recently (2006) induced Uwe-Karlsten Heye, the spokesperson of the former chancellor Gerhard Schröder, to sound the alarm: 'In Brandeburg and other German regions there are small and middle-sized towns where it is inadvisable for people with different coloured skins to go, for they might not come back alive'.

3. The Urban Conflicts in Italy

Finally, a few words on Italy. This country has not yet experienced urban conflicts connected with migration so serious as those of the European countries mentioned above. But also in Italy there have already been a lot of urban conflicts clearly due to the impact of immigration, in spite of the strange negationism of most of its leftist political figures. I quote the statement of the current president of the Republic Giorgio Napolitano, who, when he was minister of the Interior (1966), affirmed that he 'would never concede a single thing to the idea that the tensions present in some areas were due to immigration', while he had better refer to the idea that they were *only* or *mainly* due to immigration. Recently, however, even some leftist politicians have openly recognized the relationships existing between some urban conflicts and immigration. Among them, the former leader of the left coalition Romano Prodi (2005), as we have mentioned above, and his minister of the Interior Giuliano Amato (2006), who has even defined the situation existing in some quarters as 'a *banlieue* in embryo'.

The lower level of ethnic urban conflictuality that has been present up to now in Italy is due to various factors. Among them: 1) the fact that immigration is a relatively recent process; 2) the consequent low presence of second generation immigrants; 3) the low percentage of immigrants on the total population; 4) the high percentage of female immigrants since the very beginning of immigration (around 50%, though very unequally distributed among the various ethnic groups); 5) the high percentage of asylum-seekers among the immigrants, though only a very few of them are eligible as refugees, according to the Geneva Convention; 6) the scattering of immigrants throughout the country and within the towns themselves; 7) the influence exerted on Italian political culture by religious organizations, trade unions and some political parties clearly inclined to international solidarity.

On the contrary, other factors have favoured the development of these conflicts. Among them: 1) the beginning of immigration in a phase of serious economic crisis; 2) a kind of immigration due much more to the push factors in the countries of origin than to the pull factors in the country of destination; 3) the high percentage of illegal immigrants; 4) the very rapid increase in immigration; 5) the insufficient control of the territory by the police and the armed forces; 6) the extended practice of illegality, including black work, with a certain tolerance for both small and large criminal organizations, even outside their regions of origin and traditional presence.

Actually, in Italy there have been quite a lot of urban conflicts variously connected with immigration. Among them, we can single out the following types:

1) the reaction of citizens to insecurity and social and urban degradation due, or attributed, to immigration. Particularly in some central and semi-central areas of the greatest towns there have been demonstrations organized by spontaneous groups of citizens or some *ad hoc* committees. Important actions of this kind occurred in Milan, in the early '90s in the areas more largely affected by petty criminality, sale of drugs and prostitution due to immigrants (Melotti, 1993), and, later, in the 'Chinese quarter'. Similar demonstrations also occurred in Turin, Genoa and, more recently, Rome, in the Esquilino quarter, characterized by an increasing presence of the Chinese, involved in business partly controlled by their criminal organizations (Melotti, 2007);

2) the reactions, sometimes very violent, of immigrants to the clearing of some areas or some buildings that they had illegally occupied. Clashes of this kind occurred in Rome, where, in the early '90s, the building of a former pasta factory, La Pantanella, was occupied by 2,500 immigrants of various ethnic groups, and in Milan, for a farm house in Lambrate area, and, later, for some building near the Central Railway Station. Similar reactions also occurred for the clearing, or the attempted clearing, of some illegal nomad camps in Rome, Milan, Bologna and many other towns;

3) the ethnic clashes between various groups of immigrants. Bloody conflicts of this kind occurred in Rome, during the occupation of La Pantanella, and, recently, in Padua, in an area where immigrants of various ethnic groups lived together in illegal conditions and social degradation. There, after some clashes between Moroccans and Nigerians involved in drug traffic and other crimes (July 24 2006), the mayor of the town, supported by his leftist majority, decided to construct a steel barrier 3 m in height in order to isolate this area from the rest of the town. Strangely enough, this initiative was appreciated even by the minister for Social

Solidarity Paolo Ferrero, an exponent of Rifondazione Comunista, the extreme leftwing of the then government coalition, who underlined the necessity of reducing the level of conflictuality due to similar ghettos;

4) the clashes between the gangs of young immigrants or immigrants' children. The first important cases occurred in Genoa and Milan, in 2005 and 2006, between gangs of young Ecuadorians and Peruvians, who also attacked young Italians.

Some of these conflicts are clearly becoming more frequent and more serious and are spreading throughout the country. Therefore, it is necessary to pay much attention to the evolving situation, especially after the 2005 events in France. The bell of Saint-Denis rang also for Italy.

References

Augé, M. (2006) 'Il decalogo delle periferie', La Stampa, September 15, p. 27.

Baptiste, F. (1994) 'L'immigrazione a Lione: frammento di una cronaca pluridecennale', pp. 47-65, in Ires [Istituto di ricerche economiche e sociali del Piemonte], Le chiavi della città: politiche per gli immigrati a Torino e Lione, Torino: Rosenberg & Sellier.

Cohn-Bendit, D. and Schmidt, T. (1992) Heimat Babylon: Das Wagnis der multikulturellen Demokratie, Hamburg: Hoffman & Campe.

Delarue, J. M. (1991) Banlieues en difficulté. La relégation, Paris: Syros.

Delle Donne, M. and Melotti, U. (eds) (2004) Immigrazione in Europa. Strategie di inclusione - esclusione, Roma: Ediesse.

Donzelot, J. (2006) Quand la ville se défait, Paris: Seuil.

Dubet, F. (1992) Les quartiers d'exiles, Paris: Seuil.

European Commission (2000) Communication from the Commission to the Council and the European Parliament on a community immigration policy [com (2000) 757 final], Bruxelles.

Ferrarotti, F. (2005) 'Perché la più disperata è la gente di colore', Il Messaggero, September 3, pp. 1 and 11.

Gallissot, R. and Moulin, B. (eds) (1995) Les quartiers de la ségrégation: tiers-monde ou quart-monde?, Paris: Institut Maghreb-Europe et Karthala.

Lagrange, H. and Oberti, M. (eds) (2006) Émeutes et protestations: une singularité française, Paris: Presses de Science Po.

Melotti, U. (1992a) L'immigrazione: una sfida per l'Europa, Roma: Edizioni Associate.

—. (1992b) 'Migrazioni internazionali e integrazione sociale: il caso italiano e le esperienze europee', pp. 1-38, in A. Carvelli (ed.), Analisi dei bisogni e offerta di servizi per gli stranieri extracomunitari nell'area milanese, Milano: Irer [Istituto regionale di ricerca della Lombardia].

—. (1993) 'Il disagio metropolitano', pp. 157-169, in S. Allievi (ed.), L'immigrazione fra passato, presente e futuro, Milano: Iref [Istituto di ricerche educative e formative] e Comune di Milano.

—. (1995a) 'Le problème de l'intégration sociale dans les villes: spécificité du cas italien', pp. 58-108, in J. Donzelot and M. C. Jaillet (eds), Les 'zones urbaines défavorisées' : leur diagnostiques, les politiques en leur direction et la question de la justice sociale en Europe et en Amérique du Nord, vol. 1, Paris: Plan Urbain.

—. (1995b) 'Aux racines des conflits: les politiques migratoires en Europe', pp. 81-88, in R. Gallissot and B. Moulin (eds) q.v.

—. (1996) 'Immigrati e autoctoni in Italia: conflitti etnici o sociali?', Annali di Sociologia / Soziologisches Jahrbuch, n. 12: 439-454 (German transl. 'Einwanderer und Einheimische in Italien: ethnische oder soziale Konflikte?', ibi: 455-472).

—. (ed.) (2000a) Etnicità, nazionalità e cittadinanza, Roma: Seam.

—. (ed.) (2000b) L'abbaglio multiculturale, Roma: Seam.

—. (2004) Migrazioni internazionali, globalizzazione e culture politiche, Milano: Bruno Mondadori.

—. (2006) 'Immigrazione e sicurezza: quadro generale e situazione specifica con particolare riferimento al caso dei cinesi a Roma', pp. 69-188, in E. Pföstl (ed.), Multiculturalismo e sicurezza: il caso della comunità cinese a Roma, Roma: Istituto di Studi Politici S. Pio V.

—. (2006) 'Migration policies and political cultures in Europe: a changing trend', International Review of Sociology / Revue Internationale de Sociologie, 2: 191-208.

—. (2007) Le banlieues. Immigrazione e conflitti urbani in Europa, Roma: Meltemi.

Napolitano, G. (1996) 'Speech for the presentation of Dossier Immigrazione', La Repubblica, October 25, p. 2

PART TWO:

THE NEW BOUNDARIES
BETWEEN BODIES AND TECHNOLOGIES

NANO-BODY OR NOBODY? RADICAL LIFE EXTENSION OF A DISEMBODIED SELF

CELINE LAFONTAINE AND MICHELLE ROBITAILLE

Abstract: Some of the nanotechnologies' promises (such as improved physical, cognitive and sensorial aptitudes or increased life expectancy) profoundly overturn our representations of living matter, the human being, materiality and death. Symbolically, we are witness to a twofold process of the naturalization of the technical and an artificialization of nature. This chapter analyzes how nanotechnologies, especially the technological model on which they are based, transforms our way of seeing the body, which seems now like a combination of atoms to be organized, reorganized, repaired or even immortalized.

Starting with an analysis of scientific discourse, we will analyze the effects of the informational model unique to the nanotechnologies on the representation of the body, particularly in perceiving and approaching the human-machine relationship. We will see that with nanotechnologies we are no longer dealing with the familiar heuristic comparison of body and machine, but rather with a new ontology of the body *as* assortment of programmed machines. It therefore becomes imperative to create the informational human being using these technologies—adapting it to its environment—given that it is malleable and reprogrammable. In a critical perspective, we will see the consequences of this technoscientific vision of the body in the representation of subjectivity.

The fulgurant acceleration of technoscientific advances has profoundly and permanently transformed our social, cultural and symbolic paradigms. Whether we are talking about identity issues connected with the redefinition of human/machine boundaries or new social practices stemming from cyberculture, the technosciences are at the heart of contemporary transformations. One of the technoscientific developments of postmodern society, nanotechnologies now transform inert or living matter through molecular docking by creating new matter, the physical, chemical and biological properties of which are still unknown. Presented as the conquest of the 'infinitely small,' nanotechnologies promise nothing less than the ability to manipulate matter on an atomic scale.

As much in their design as in their potential application, nanotechnologies participate in a logic of crossing the living with the non-living, the natural with the artificial and human beings with machines. As Jean-Pierre Dupuy (2002) and Bernadette Bensaude-Vincent (2004a; 2004b) rightly point out, nanotechnologies are the outcome of the informational paradigm. The huge possibilities opened up by the conquest of the infinitely small have let nanotechnology researchers' imaginations run wild. Coupling living organisms and inert matter at the molecular level has allowed for an ever-expanding conception of the limits of the human body, for example, using silicon chips and nanorobots (Pautrat, 2002). Certain researchers, such as Ray Kurzweil (2005), are already talking about a new species improved through genetic engineering and nanotechnologies. In fact, the unequalled potential of the nanotechnologies feeds some scientists' hope of radically transforming and improving human nature by fusing humans and machines.

Nanotechnologies participate in the twofold process of making nature artificial and technologies natural (Bensaude-Vincent, 2004). In keeping with the informational paradigm, this twofold process is based on a broad and adaptive conception of the human body. Indeed, as we will see, the representations of the body that accompany the development of nanotechnologies correspond perfectly to the theoretical assumptions on which the post-human concept, as defined by Katherine Hayles (1999), is based. From the post-human point of view, human beings are by nature essentially informational, and their biological bodies are an evolutional accident. In fact, the body is seen as a prosthesis that can be modified and controlled so as to fuse human and machine together (Hayles, 1999: 2). Epistemologically and historically, the post-human's imagination, as presented by nanotechnologies, cannot be disassociated from the technoscientific revolution that was sparked by cybernetics at the end of the Second World War.

From Cyber to Nano

Defined as the *science of control and communication in animals and machines*, cybernetics is the prototype of technoscience; in other words, a cognition mode focused on operational and technical control (Wiener, 1988). Briefly reviewing its initial project indeed allows us to better understand the current convergence of nanotechnologies, biotechnologies, information technologies and cognitive sciences, or what has come to be commonly known as NBIC (nano, bio, info, cogno). Far from a socio-historical accident, this convergence is rather the determining character of the informational paradigm in the development of technosciences that started in the 50s.

Historically, cybernetics takes root at the core of the technoscientific project implemented by the American government during the Second World War. Norbert Wiener, considered as one of the founders of cybernetics, participated in the war effort by devising a servomechanical shooting device, the AA predictor. In an article entitled 'The Ontology of the Enemy'(1994), science historian Peter Galison demonstrated the significance of this military experiment as a defining moment in the elaboration of the cybernetic model. In his article, he states that engineering an artillery system capable of following and identifying its target effectively is what inspired Wiener to develop a theoretical model in which the pilot is integrated as a part of a self-regulated machine. Based on the feedback notion, the analytical model he developed during this period stems, in fact, from a conceptual absence of differentiation between human and machine. The pilot represents an integral part of the technical device. In fact, the enemy pilot is the first-ever cyborg model created and later becomes the icon of the cybernetic subject after the war. It is in that sense that Peter Galison uses the phrase 'Ontology of the Enemy' (1994).

Based on his AA predictor work, Wiener and his colleagues began an epistemological revolution by rejecting the intrinsic study of beings and things, and focusing the analysis instead on the interactions between objects regardless of their nature (physical, biological, artificial or human). This is clearly illustrated in a text he co-signed with Bigelow and Rosenblueth in 1943, in which the ontological difference between humans and machines is replaced by a hierarchical classification of behaviour, based on a behaviourist model, where teleological behaviour dominates (Wiener et al., 1943). Feedback machines are thus promoted to the rank of complex-intelligence entities, alongside human beings.

As sociologist Philippe Breton (1995) has demonstrated, Wiener elevated entropy to the rank of metaphysical truth. In the wake of the war, it is assimilated to the chaos, disinformation and disorganization that threaten the social order. Information, regarded as a negentropical principle, can temporarily fight this force that triggers apathy and destruction (Breton, 1995: 33). Equally as abstract as the concept of energy, the notion of information then becomes a principle of statistical quantification whose universal scope is equalled only by its indifference toward the specific nature of signals (physical, biological, technical or human) (Dion, 1997). Formulated simultaneously by Claude Shannon and Norbert Wiener in 1948, the theory of information sees an unparalleled level of diffusion in the scientific community (Segal, 2003). Understood in terms of information exchange, communication becomes the source of any organization.

By defining humans based only on the degree of complexity of their intelligence, the father of cybernetics implies and proclaims that the ontological value of a machine would be identical to that of a human being:

> It is in my opinion, therefore, best to avoid all question-begging epithets such as 'life', 'soul', 'vitalism', and the like, and say merely in connections with machines that there is no reason why they may not resemble human beings in representing pockets of decreasing entropy in a framework in which the large entropy tends to increase (Wiener, 1988: 32).

Many scientific historians, particularly Lily Kay and Evelyn Fox Keller, have shown that the ascendance of cybernetics over the development of molecular biology and genetic engineering is first and foremost metaphorical (Fox Keller, 1999; Kay, 2000). However, the performative power of cybernetic metaphors brings up social, cultural and environmental issues that are very real indeed. It is nevertheless with nanotechnologies that the technoscientific revolution put in motion by cybernetics takes on its full meaning. Nanotechnologies therefore fall into line with Wiener's cybernetics project, which, in his book *Humans' Use of Human Beings* already affirmed the following: 'We have modified our environment so radically that we must now modify ourselves in order to exist in this new environment. We can no longer live in the old one' (Wiener, 1988: 56).

Nanotechnologies in Pursuit of New Frontiers

Product of a technoscientific convergence of quantum physics, chemistry, microelectronics, computer science and genetic engineering, nanotechnologies are characterized by manipulation and recombination of matter at the atomic level. The issue of dimension is now so fundamental that some authors use terms such as 'nanocosm,' 'nanoworld' or 'nanoculture' to indicate this changed scale by which we conceive of and manipulate matter (Pautrat, 2002; Atkinson, 2003; Los Angeles Contry Museum of Art et al., 2004). Historically, it was physicist and Noble Prize winner Richard Feynman who was the first to support the idea of a reorganization of matter on the atomic level during a famous lecture given by the American Physical Association in 1959 (Feynman, 1959). However, it was only with the publication in 1986 of *The Engines of Creation,* a book by engineer Eric Drexler, that a radical shift in perspective occurred in the technoscientific fields (1986). Considered a visionary, Drexler proposed a view of the future in which nanorobots are called upon to fuse living species and machines by manipulating matter from the bottom up, atom by atom. Yet, we must emphasize that the bottom-up model developed by Eric Drexler is directly influenced by the theory of self-regulating mechanisms developed by John Von Neumann in 1948 (Dupuy, 2002). Taking Wiener's cybernetics one step further, the Von Neumann model points towards a second cybernetics and towards complexity and the theory of self-regulated systems (Dupuy, 2002).

The notions of control and manipulation of matter that we come across quite frequently in public debate over nanotechnologies demonstrate the epistemological predominance granted to technical application. As Jan C. Schmidt (2004) points out, these logics participate in a technological reductionism that contributes to blurring the boundaries between science and technology. Nanotechnology, indeed, has become an 'umbrella' term designating all technoscientific mutations. Thus, as Chris Hables Gray supports in his *Cyborg Citizen*: 'nanotechnology is not just a little corner of contemporary science/engineering/business: it is the expression of our postmodern age, replete with the postmodern characteristics: bricollage, the centrality of speed and information' (Gray, 2002: 182).

A direct extension of the informational paradigm, nanotechnologies are based on the 'philosophical plan,' which can be analyzed through three fundamental dimensions (Dupuy, 2002). The first of these dimensions supposes that everything is algorithmic and an informational machine: mind, body, cosmos, nature and life. The algorithm analogy in nanotechnology has exceptional scope as it involves a artificial treatment of nature in its entirety. The second dimension is the emergence of a new social actor: the nanoengineer. The engineer's role in the 'manufacturing' of nanotechnologies cannot be summarized as one who designs structures, but rather as one who facilitates the evolutionary process: 'in this case, far from seeking mastery, the engineer knowingly and deliberately plays the magician's apprentice: he judges his undertaking as all the more successful if the machine he configures takes him by surprise' (Dupuy, 2002: 10). Therefore, it is no longer a question of subjugating nature, but rather of fabricating it and thus creating life. The philosophical plan's third dimension concerns the evolution implicit in an inevitable informational process that humans are obliged to participate in, notably through technological development. In these ways, nanotechnologies participate in the twofold process of making informational systems natural and nature artificial.

By making the convergence of technosciences possible (in particular through the NIBC National Science Foundation (Roco and Bainbridge (2002)), nanotechnologies give the impression of being able to make good on practically all scientific promises. Among these numerous promises, the possibility of modifying and improving the body takes a central place. In fact, as we will see, nanotechnologies participate in a redefinition of being human, starting with a new relationship between nature and artifice, and a technoscientific evolutionism incarnated by the metaphorical figure of the post-human.

Technologization of the Body: the Nano-Body

The nanotechnologies encompass representations of the body and the discourse that accompanies their development has both descriptive and normative dimensions. To fully understand this issue, it is essential to refer to the view certain researchers hold regarding the body and its limits.

The idea of placing millions of autonomous nanorobots inside one's body might seem odd, even alarming. But the fact is that the body already teems with a vast number of mobile nanodevices, built not by human hands but by nature. Consider neutrophils, lymphocytes and other white blood cells. By nanoscale standards, they are quite large, measuring some 10,000 nanometers across, but they function as natural nanorobots, constantly roving about the body, repairing damaged tissues, attacking and eating invading micro-organisms, and sweeping up foreign particles for various organs to break down or excrete. All of us are utterly dependent on those cells for survival (Freitas Jr., 2000: online).

Using nanotechnology, we can design fully intelligent polymorphic material that consists, like your body, of trillions of microscopic machines. Like your cells, each machine will have a substantial local program and information storage, but will act in accordance with patterns of global information. Unlike your cells, they will be more quickly and more widely reprogrammable, adopt a wider array of functions, and look like spiders rather than jellyfish. (Storrs Hall, 1993: online)

Physical disorders stem from the disorganization of atoms; reparative machines will be able to return health since they will be able to reorganize atoms into a functional order. (Drexler, 1986: 121)

Seen from the perspective of researchers in the nanotechnologies, the body appears to be the product of a natural assembly of already existing nanorobots. Building on a model inspired by biology, nanorobots acquire a natural status and imbed themselves into the deepest parts of the human body, which is already defined by its 'molecular nanomachine' make-up.

A comparison of the body and the machine is no longer talked about; rather, the new ontology views the body as an assortment of programmed machines. The figure of the cyborg—as product of a perfect fusion of the organic and the cybernetic—is no longer the avatar of science fiction but, rather, it is being gradually anchored into reality (Gray, 2002). This paves the way for the idea of *re*-programming, a notion we'll come back to shortly.

Many researchers see the biological body as being outdated. To them, nanotechnologies are going to perfect the body to make it better adapted to the 'society of information' and therefore perform better. We are not talking about perfecting in the classic sense, but rather in the technoscientific sense (Winner, 2002). For example, consider the following excerpt, in which Ray Kurzweil deplores the inefficiency of our digestive systems, and then another by Ted Sargent on rewriting DNA:

> Today, this biological strategy (the digestive system) is extremely counterproductive. Our outdated metabolic programming underlies our contemporary epidemic of obesity and fuels pathological processes of degenerative disease such as coronary artery disease, and type II diabetes. (Kurzweil, 2003)

> Inside we are abuzz with molecular mechanics [...] The machine inside ours cells that manufacture protein from DNA instructions can make mistakes...(Sargent, 2005: 33-34).

Even biological memory is no longer adequate, explains E. Garcia-Rill:

> Once we added symbols, alphabets, and mathematics, biological memory became inadequate for storing our collective knowledge. That is, the human mind became a 'hybrid' structure built from vestiges of earlier biological stages, new evolutionarily-driven modules, and external (cultural 'peripherals') symbolic memory devices (books, computers, etc.), which, in turn, have altered its organization, the way we 'think' (Donald, 1991) (In Roco and Bainbridge (2002): 228).

The idea of humans' technoscientific perfectibility is very explicit in the following excerpt from the NBIC report published by the National Sciences Foundation:

> Examples of payoffs may include improving work efficiency and learning, enhancing individual sensory and cognitive capabilities, revolutionary changes in healthcare, improving both individual and group creativity, highly effective communication techniques including brain-to-brain interaction, perfecting human machine interfaces including neuromorphic

engineering, sustainable and 'intelligent' environments including neuro-ergonomics, enhancing human capabilities for defense purposes, reaching sustainable development using NBIC tools, and ameliorating the physical and cognitive decline that is common to the aging mind (Roco and Bainbridge (2002): ix).

As we have already pointed out, the technoscientific representation of the body as *outdated*, *perfectible* and *informational* participates in a technological determinism. This is rooted in the very foundation of the plan to improve and increase the abilities of the human body, which is supported by many researchers in nanotechnologies. From their point of view, it becomes imperative to model the human being using these nanotechnologies—thus adapting humans to their environment—given that it is inefficient and reprogrammable. This type of representation of the body clearly advocates a legitimization of technoscientific research.

In *More than Human* (2005), Ramez Naam exposes certain arguments in favour of the plan to improve the human body. Among the arguments presented, let us retain the idea that the improvement of the human body is in line with an increased individual liberty. This argument is found in the theory on nanotechnologies:

> Should individuals and families have the right to alter their own minds and bodies, or should that power be held by the state? In a democratic society, it's every man and woman who should determine such things, not the state...Governments are instituted to secure individual rights, not to restrict them (Naam, 2005: 6-9).

> The right of each individual to use new knowledge and technologies in order to achieve personal goals, as well as the right to privacy and choice, are at the core of the envisioned developments (Roco and Bainbridge (2002): x).

This argument is also at the heart of *Citizen Cyborg*, a book by James Hughes, Executive Director of the World Transhumanist Association: this book argues that transhuman technologies, technologies that push the boundaries of humanness, can radically improve our quality of life, and that we have a fundamental right to use them to control our bodies and minds (Hughes, 2004: xii).

Another argument, brought up by Ramez Naam, addresses the so-called natural tendency that humans have always had to want to improve themselves:

> Far from being unnatural, the drive to alter and improve on ourselves is a
> fundamental part of who we humans are. As a species we've always looked
> for ways to be faster, stronger, and smarter and to live longer (2005: 9).

We will not discuss these arguments further here—our goal was simply to
show that experts present the project to improve the human body not only
as inevitable but as being entirely humanist and democratic. Do we have a
sense today of how nanotechnologies should be used to improve the body?

Re-programming the Body

The philosophical plan underlying nanotechnologies also has normative
and prescriptive impacts: 'It would seem that, for the very first time, we
could put into practice a normative definition of humanity and decide
which are the characteristics that determine what it is to be human'
(Maestrutti, 2006). Therein, we can consider a contemporary form of bio-
power: a form of diffuse domination that technoscientific knowledge
exercises over individuals who subject themselves thereto (Foucault,
1976). According to Foucault, confronted with new scientific knowledge,
individuals develop standards, institutionalize them and conform to them.
Moreover, in a society where performance and pushing one's personal
limits are not only rewarded but already the norm, individuals are
increasingly inclined to model themselves using, for example, plastic
surgery, body building, psychotropic medications and diets (Lypoveysky,
1983; Le Breton, 1999; Lasch, 2000). This paves the way for even more
radical modifications in the technosciences, such as implanting silicon
chips in the brain and heightening the senses (Featherstone and Burrows
(1995).

The ideal of perfectibility that the nanotechnologies advocate is
illustrated by the accomplishments these technologies promise in the
medical field (early detection of cancers, medications that target sick cells,
insulin control, etc.). In his article 'Human Body Version 2.0,' Ray
Kurzweil proposes, for example, to improve health using nanorobots
programmed to replace the red blood cells and improve digestion:

> One of Freitas' designs is to replace (or augment) our red blood cells with
> artificial 'respirocytes' that would enable us to hold our breath for four
> hours or do a top-speed sprint for 15 minutes without taking a breath
> (2003).

> In an intermediate phase, nanobots in the digestive tract and bloodstream
> will intelligently extract the precise nutrients we need, call for needed

additional nutrients and supplements through our personal wireless local area network, and send the rest of the food we eat on its way to be passed through for elimination (2003).

Loyal to cybernetics' military and informational paradigm origins, the plan to re-programme the body using the nanotechnologies is intended to enable soldiers to adapt to extreme conditions:

> Without the use of drugs, the union of nanotechnology and biotechnology may be able to modify human biochemistry to compensate for sleep deprivation and diminished alertness, to enhance physical and psychological , and to enhance survivability rates from physical injury (Roco and Bainbridge (2002): 329).

Senses Programmed to be Decoded

Half human, half machine, the cyborg incarnates the ideal of a being whose performance is improved through electronic or genetic prostheses. Feeding the nanotechnological imagination, the figure of the cyborg was, as we know, already planted as a seed in cybernetics. Let's not forget that, at the end of his life, Norbert Wiener dedicated much of his research to prosthetics. (Lafontaine, 2004: 166). It is certainly not an accident then that the extension of the senses and their improved performance are focal points of nanotechnologies. Pushed further, this research leads to the production of prostheses that modify the perception of the physical and social environment:

> The next decade will see great strides in personal wearable technologies that enhance people's ability to sense their environment. This sensing will focus on at least two different areas:
> a) *social sensing*, in which we may augment our ability to be aware of people in our immediate vicinity with whom we may wish to connect (or possibly avoid!)
> b)*environmental sensing*, in which we may augment our ability to sense aspects of our environment (for example, the quality of the air we are breathing) that may be hazardous to us but that our normal senses cannot detect (Roco and Bainbridge (2002): 164).

Similarly, it might become possible to filter certain stimuli:

> By monitoring our sensory gating capability, our ability to appraise and filter out unwanted stimuli can be assessed, and the chances of successful *subsequent* task performance can be determined (Roco and Bainbridge (2002): 229).

The sensorial stimuli considered to be information that our senses 'capture' and that our brains 'decode' will perhaps be changed, accentuated or even concealed. Might we even eventually eliminate all perception of suffering, fatigue or sadness, as philosopher David Pearce suggests in his manifesto 'The Hedonistic Imperative'? Extrapolated to an extreme, this conception leads to the idea of the obsolescence of biological senses in the world of virtual reality:

> One application will be to provide full-immersion virtual reality that encompasses all of our senses. When we want to enter a virtual-reality environment, the nanobots will replace the signals from our real senses with the signals that our brain would receive if we were actually in the virtual environment (Kurzweil, 2003).

From this point of view, the senses are absolutely unnecessary to being human, to humans' understanding of the world and their ability to act; or, in other words, 'The Senses Have No Future' as suggests the title of an article by Hans Moravec, a robotics researcher:

> It would be far better to bypass all the sensory processing, and insert the message from the computer directly into the thinking portions of your brain. In such manner all our senses will become obsolete, as our physical environment is inexorably refined from a rough physical place into a densely interconnected cyberspace (Moravec, 1997: 1).

This is directly connected with the next dimension of the representation of the technoscientific body that we will examine: the brain-to-computer interface.

Brain-to-Computer and Brain-to-Brain Interfaces

The desires to improve intellectual performance, to increase memory, to accelerate learning and to be able to access the Internet at all times are ubiquitous in the discussion surrounding nanotechnology research. In this view, the intellectual processes are essentially designed as informational exchanges. Some go so far as to say that the 'most valuable sixth sense for our species would be a sense that would allow us to quickly understand, in one big sensory gulp, vast quantities of written information (or even better, information encoded in other people's neural nets)' (Roco and Bainbridge (2002): 109). The predominant influence of cybernetics and the informational paradigm is demonstrated by the brain's assimilation into a machine: 'What is the data structure for human memory? Where are the bits? What is the capacity of the human memory in gigabytes (or petabytes system)?' (Roco and Bainbridge (2002): 168), 'What kind of software does the brain use?' (Roco and Bainbridge (2002): 228). Once it is understood as a computer, the brain can be connected to other computers, biological or not:

> The most important application of circa-2030 nanobots will be to literally expand our minds. We're limited today to a mere hundred trillion interneuronal connections; we will be able to augment these by adding virtual connections via nanobot communication. This will provide us with the opportunity to vastly expand our pattern recognition abilities, memories, and overall thinking capacity as well as directly interface with powerful forms of nonbiological intelligence. (Kurzweil, 2003: online)

Indeed, the idea of one sole brain and a collective intelligence is expressed here: everyone has access at all times to others' ideas and electronic resources (Lafontaine, 2004). Once conceived of as information, the brain (or rather its contents) could be transferred into a computer, making the biological body perfectly obsolete:

> Uploading a human brain means scanning all salient details and then reinstantiating those details into a suitably powerful computational substrate. This process would capture a person's entire personality, memory, skills and history (Kurzweil, 2005: 198-9).

The informational model underlying this plan to transform and improve the body through nanotechnologies thus leads—paradoxically—to a rejection and negation of the body. The body is thus desubjectified while the object is literally disembodied. And therein we witness a double disappearance.

The Technologization of Nature and Evolution

In conjunction with the informational paradigm, the nanotechnologies' basic proposition of unifying material on the atomic level leads to the theoretical dissolution of body and subject:

> No, Greta [Garbo] was not gone. It was just that her atoms were in all the wrong places, spread around the earth, but still somewhere in the material world...Her very sensory endowment was archived, and in the air were suspended the very elements which, appropriately arranged, had given her life (Sargent, 2005: x).

We should note that, in the nanotechnologies, the informational machine analogy holds sway for both those who subscribe to the Drexler hypotheses, as well as for its detractors. Indeed, many researchers in the nanotechnologies, notably Richard Smalley, have been sceptical of Drexler's hypotheses about self-replicating nanorobots (Bensaude-Vincent, 2004a; 2004b). However, all base their ideas on organic systems and then go on to construct—and validate—their mechanical model of living things.

According to Drexler's 'molecular manufacturing' model, nature is made up of nanorobots that humans will eventually be able to reproduce and imitate (1986). For researcher Marvin Minsky, life itself proves the feasibility of such nanorobots:

> It seems quite strange for anyone to argue that you cannot build powerful (but microscopic) machinery – considering that our very own cells prove that such machines can indeed exist. And then if you look inside your cell you will find smaller machines that cause disease. Most arguments against nanotechnologies are arguments against life itself' (cité dans Bensaude-Vincent, 2004).

In addition to manipulating molecules already existing in nature, the nanotechnologies propose to create new ones, which raises questions about the relationship between nature and artifice. 'Is it at all possible to distinguish between nature and technology if nature has already become technologically malleable at the level of molecules? Can nature- if it is distinguishable from technology at all – set limits to technology?' (Schiemann, 2005). This brings us to the question of how the nanotechnologies make nature artificial and artifice natural (Dupuy, 2002). In addition to blurring the boundaries between the natural and the artificial, this makes any ethical reflection difficult because, seen from this angle, engineers seem simply to reproduce in laboratories what nature has done since the beginning of time:

> By building the artificial, observes Postrel, we do not overthrow nature, but cooperate with it, using nature's own art to create new natural forms. Our artifice alters the path of nature, but it does not end it, for nature has no stopping point, no final shape. It is a process, not an end (Freitas Jr., 1999: online).

When the subject of nature is brought up in discussions about the development of nanotechnologies, it is generally in relation to the evolutionary process, which is deemed to be too slow. And therefore this is a process in which humans must inevitably take part, via new NBIC engineers. Some researchers in the nanotechnologies draw a parallel between biological evolution and 'Moore's law,' which stipulates that, thanks to miniaturization, the number of transistors on a silicon chip (and therefore the resulting power of the processor) doubles every 18 months. According to Ray Kurzweil and Hans Moravec, this exponential growth of computer power leads us to think that computers will soon be just as powerful as the human brain and will eventually outdo it (Hughes, 2004). Compared to the rapid growth of computer power, biology seems like a dangerously slow process, thus inspiring a sometimes imperious tone in discourse: 'The success of this convergent technologies priority area is essential to the future of humanity' (Roco and Bainbridge (2002): xiii).

On the Road to Immortality

The theoretical conceptions on which developments in the nanotechnologies are based have metaphysical implications, as Collin Milburn explains: 'The possible parameters of human subjectivities and human bodies, the limits of somatic existence, are transformed by the invisible machinations of nanotechnology—both the nanowriting of today and the nanoengineering of the future—facilitating the eclipse of man and drawing of the post-human condition' (Milburn, 2004: 114). The informational view of the human being, as is expressed in discourse surrounding post-humans and the nanotechnologies, takes for granted that corporality will be outdone and the boundaries of biological materiality negated. Desubjectified and disembodied, informational beings can at last aspire to earthly immortality—the infinite extension of their 'existence'—by transposing their brains to another material support. Since the nanotechnologies promise to control both the structure of the matter and the structure of the biological body, they present themselves as a kind of universal and omnipotent solution to the 'problem' of mortality. Thus, researchers from various disciplines –AI, cryogenics, biogerontology, prosthetics—are counting on nanotechnologies to overcome obstacles in their projects to prolong life.

Closer to science fiction than to real technical possibilities, some researchers do not hide their ultimate fantasy: beating death. One of the primary promises of nanomedicine is to increase life expectancy until it is able to stop and then reverse the aging process, which is perceived as an illness that can be 'treated':

> Most investigators think aging is the result of a number of interrelated molecular processes and malfunctions in cells. Thus if nanomedicine can learn to reverse most cellular malfunctions, middle-aged and even elderly

people should be able to regain most of their youthful health, strength and beauty, and to enjoy an almost indefinite extension of life (Freitas Jr., 2000: online).

Nanodoctor Freitas Jr. offers a three-step 'dechronification' or 'rolling back the clock' procedure requiring nanotechnological devices. First, empty each cell of accumulated toxins, then replace those chromosomes showing genetic errors and, finally, locate one by one the cases of more serious damage to cell structure (Freitas Jr., 2003). In the same vein, the book *Fantastic Voyage* (2005) by Ray Kurzweil and Terry Grossman addresses the reasons behind the body's deterioration in life:

> We are beginning to understand aging, not as a single inexorable progression but as a group of related biological processes. Strategies for reversing each of these aging progressions using different combinations of biotechnology techniques are emerging (Kurzweil and Grossman (2005): 4).

For Aubrey de Grey, biogerontologist in the genetics department at Cambridge University, there is no doubt that a radical extension of life is imminent: 'All the knowledge needed to develop *engineered negligible senescence* is already in our possession – it mainly just needs to be pieced together' (in Kurzweil and Grossman (2005): 24). He has identified seven key elements that are at the source of aging and that accelerate cells' endogenous degenerative process, which leads to death. For each of these processes (genetic mutation, cellular intoxication, cellular atrophy), the author proposes a strategy for stopping or reversing its effects.

Like Drexler, an author who dedicated a whole chapter of his *The Engines of Creation* to cryogenics, cyronicists are counting on the nanotechnologies to not only repair cells one by one, but to bring them back to life (Regis, 1990). Made possible by the nanotechnologies, cryogenics purports to suspend life in order to defy death. Note that patients are not considered dead but rather 'potentially living,' 'suspended' or in 'biological coma' (Regis, 1990: 128-9).

Seen as a purely informational process, the body now has virtually no limits. Thus, the philosophical plan underlying certain research in the nanotechnologies aims not only to improve humans' adaptation and performance, but also to make them literally all-powerful and omnipotent. We might say that the post-human ideal has the ambition to modify the biological human body and even goes so far as to remove it from the picture altogether.

The Post-human: a Disembodied and Desubjectified Being

Direct inheritors of cybernetics and the informational paradigm, the nanotechnologies intersect with numerous contemporary technoscientific advances. Erasing the boundaries between science and science fiction, the discourses that accompany the development of nanotechnologies entail a complete redefinition of human beings in terms of technological improvement and perfection (Milburn, 2004). With his improved adaptation and performance, and virtual immortality, what the post-human loses in subjectivity, he gains in potential. Perceived as a link in the informational chain, to which he must necessarily adapt, the post-human's freedom is limited to his technical and biological abilities. Potentially immortal, he incarnates the ideal of a society that is defined by network mobility and information fluctuations. Thus, the representation of the human body that underlies the development of nanotechnologies is perfectly aligned with the informational logic that is inherent to post-modern society (Lafontaine, 2004). Above and beyond the many ethical questions raised by this research intending to improve and modify the human body using nanotechnologies, we must realize that one of the primary issues in this research is the connection between corporality and subjectivity which, since the beginning of time, has been impossible to separate from our identity.

References

Atkinson, W. (2003) Nanocosm. The Big Change That's Coming from the Very Small. Canada: Viking.

Bensaude-Vincent, B. (2004) Se Libérer de la Matière? Fantasme autour des Nouvelles Technologies. Paris: INRA.

—. (2004) 'Two Cultures of Nanotechnology?', HYLE, International Journal for Philosophy of Chemistry 10(2): 65-82.

Breton, P. (1995) À l'Image de l'Homme: du Golem aux Créatures Virtuelles. Paris: Seuil.

Dion, E. (1997) Invitation à la Théorie de l'Information. Paris: Seuil.

Drexler, E. (1986) The Engines of Creation: the Coming Era of Nanotechnology. New York: Anchor Books.

Dupuy, J.-P. (2002) 'Impact du Développement Futur des Nanotechnologies sur l'Économie, la Société, la Culture et les Conditions de la Paix Mondiale. Projet de mission', Paris: Conseil général des Mines.

Featherstone, M. and R. Burrows (1995) 'Cultures of Technological Embodiment: An Introduction', Body & Society 1(3-4): 1-19.

Feynman, R. P. (1959) 'There's Plenty of Room at the Bottom', American Physical Society conference, Caltech (California Institute of Technology)

Foucault, M. (1976) Histoire de la Sexualité. Paris: Gallimard.

Fox Keller, E. (1999) Le Rôle des Métaphores dans les Progrès de la Biologie. Le Plessis-Robinson: Institut Synthélabo pour le progrès de la connaissance.

Freitas Jr, R. A. (1999) Nanomedicine, Volume I: Basic Capabilities. Georgetown: Landes, URL (consulted May 2006): http://www.kurzweilai.net/meme/frame.html?main=/articles/art0602.html

—. (2000) 'Say 'AH!'' The Sciences online (consulted May 2006): http://www.kurzweilai.net/meme/frame.html?main=/articles/art0220.html

—. (2003) 'Death is an Outrage', URL (consulted May 2006): http://www.kurzweilai.net/meme/frame.html?main=/articles/art0536.html

Galison, P. (1994) 'The Ontology of the Enemy: Norbert Wiener and the Cybernetic Vision' Critical Inquiry 21(1): 228-266.

Gray, C. H. (2002) Cyborg Citizen: Politics in the Posthuman Age. New York; London: Routledge.

Hayles, N. K. (1999) How we Became Posthuman. Virtual Bodies in Cybernetics, Literature, and Informatics. Chicago; London: University of Chicago Press.

Hughes, J. (2004) Citizen Cyborg: Why Democratic Societies Must Respond to the Redesigned Human of the Future. Cambridge: Westview.

Kay, L. E. (2000) Who Wrote the Book of Life? A History of the Genetic Code. Stanford: Stanford University Press.

Kurzweil, R. (2003) 'Human body version 2.0.', URL (consulted May 2006): http://www.kurzweilai.net/meme/frame.html?main=/articles/art0551.html

—. (2005) The Singularity is Near. New York: Viking.

Kurzweil, R. and T. Grossman (2005) Fantastic Voyage: Live Long Enough to Live Forever. London: Rodale.

Lafontaine, C. (2004) L'Empire Cybernétique: des Machines à Penser à la Pensée Machine. Paris: Seuil.

Lasch, C. (2000) La Culture de Narcissisme. Paris: Climats.

Le Breton, D. (1999) L'Adieu au Corps. Paris: Métailié.

Los Angeles County Museum of Art et al. (eds) (2004) Nanoculture: Implications of the New Technosciences. Bristol; Portland: Intellect Books.

Lypovetsky, G. (1983) L'Ère du Vide. Essai sur l'Individualisme Contemporain. Paris: Gallimard.

Maestrutti, M. (2006) 'La Singularité Technologique: un Chemin vers le Posthumain?', Vivant online, 02/03 2006, URL (consulted May 2006): http://www.vivantinfo.com/index.php?id=141.

Milburn, C. (2004) Nanotechnology in the Age of Posthumain Engineering: Science-Fiction as Science, pp. 109-129 in. Los Angeles County Museum of Art and al. (eds) Nanoculture: Implications of the New Technosciences. Bristol; Portland: Intellect Books.

Moravec, H. (1997) 'The Senses Have no Future', URL (consulted May 2006): http://www.kurzweilai.net/meme/frame.html?main=/articles/art0185.html

Naam, R. (2005) More than Human: Embracing the Promise of Biological Enhancement. New York: Broadway Books.

Pautrat, J.-L. (2002) Demain le Nanomonde. Paris: Fayard.

Pearce, D. 'The Hedonistic Imperative', URL (consulted May 2006): http://www.hedweb.com/hedethic/hedonist.htm.

Regis, E. (1990) Great Mambo Chicken and the Transhuman Condition. New York: Basic Books.

Roco, M. C. and W. S. Bainbridge (2002) Converging Technologies for Improving Human Performance. Washington: National Science Foundation.

Sargent, T. (2005) The Dance of Molecules. How nanotechnology is Changing our Lives. Canada: Viking.

Schiemann, G. (2005) 'Nanotechnology and Nature. On two Criteria for Understanding their Relationship', HYLE, International Journal for Philosophy of Chemistry 11(1): 77-96.

Schmidt, J. C. (2004) 'Unbounded Technologies: Working Through Technological Reductionism of Nanotechnology', pp. 35-50 in Baird and al. (eds) Discovering the Nanoscale. Amsterdam: IOS Press.

Segal, J. (2003) Le Zéro et le Un. Histoire de la Notion Scientifique d'Information au 20e Siècle. Paris: Syllepse.

Storrs Hall, J. (1993) 'Utility Fog: The Stuff that Dreams Are Made Of', URL (consulted May 2006): http://www.kurzweilai.net/meme/frame.html?main=/articles/art0220.html

Wiener, N. (1988) The Human Use of Human Beings : Cybernetics and Society. New York: Da Capo.

Wiener, N. and al. (1943) 'Behavior, Purpose and Teleology' Philosophy of Science 10(1): 18-24.

Winner, L. (2002) 'Are Humans Obsolete?' Hedgehog Review 4(3): 25-44.

MIRROR, MIRROR ON THE WALL, WHO IS THE .CO.ZA-TRIBE AFTER ALL?

AMELIA RICHARDS

Abstract: This chapter will aim to address this critical question by exploring the important role cyberspace plays during the process of transforming a culture of tradition and conservatism to a new and modern .co.za culture. After a brief discussion of the relevant literature and research method, the results based on chat-room interaction will focus on:

- The .co.za tribal members
- Their voice i.e. the 'co.za dictionary'
- Their lifestyles in online living spaces
- Norms and values of the .co.za-culture.

The research results will illustrate how Internet chat-rooms are constructive creative products used by South Africans as a communication tools to:

- Adapt to a rapidly changing technological environment on a global basis
- Keep abreast with the world and its totality
- Close the gap between virtual and real spaces / living
- Revisit traditional cultural belief systems, norms and values and replace them with new cyberspace cultural norms and values.

The chapter will be closed with a discussion on the Internet as a social research tool with the emphasis on qualitative research and its usability within the South African context.

1. Introduction

This chapter will aim to address this critical question by exploring the important role cyberspace plays during the process of transforming a culture of tradition and conservatism to a new and modern .co.za culture. After a brief discussion of the relevant literature and research method, the results based on chat-room interaction will focus on:

- The .co.za tribal members
- Their voice i.e. the 'co.za dictionary'
- Their lifestyles in online living spaces
- Norms and values of the .co.za-culture.

The research results will illustrate how Internet chat-rooms are constructive creative products used by South Africans as a communication tools to:

- Adapt to a rapidly changing technological environment on a global basis
- Keep abreast with the world and its totality
- Close the gap between virtual and real spaces / living
- Revisit traditional cultural belief systems, norms and values and replace them with new cyberspace cultural norms and values.

The chapter will be closed with a discussion on the Internet as a social research tool with the emphasis on qualitative research and its usability within the South African context.

2. Theory

2.1 Carl Roger's Theory of Creativity

Whilst website designers aim to create attractive websites, cyber psychologists focus on the relationship between cyberspace and the rest of our lives. These two creative symbiotic cyber-processes will be described by applying Roger's theory of creativity. Primarily this section is based on 1952 conference proceedings titled 'Toward a theory of creativity' delivered by Carl Rogers in Granville, Ohio. The value of his work lies in the fact that more than 50 years later, we can still apply his views to the development of the Internet by bringing truth to the assumption that the creative act has an universal nature and can be applied to new ideas and technological creations irrespective of the time frame.

Many contemporary authors, especially within the South African context, will agree with Rogers when arguing that the past 52 years can be described as a period where creativity was not seen as the most important point on society's agenda. Education systems produced conformists, stereotypes and individuals with an education that lacked free, creative, original thinking (Rogers 1952: 249). In the field of media development, television was described as a way of providing passive entertainment to the so-called 'coach potato' audience. In the industrial world the creative process was only available to a selected few in managerial positions, designers or executives in research departments. In summary 'in the clothes we wear, the food we eat, the books we read, and the ideas we hold, there is a strong tendency toward conformity, toward stereotypes. To be original, or different, is felt to be dangerous' (Rogers 1952: 249). At then there was the Internet...

During the 1950's, Rogers predicted 'unless man can make new original adaptations to his environment as rapidly as his science can change the environment, our culture will perish. Not only individual maladjustment and group tension, but international annihilation will be the price we pay for a lack of creativity' (Rogers 1952: 250). When one looks at the generation differences one will see that the cyberspace culture has developed because of a need of individuals to control their environments not looking toward a governing body to make those decisions for them. What makes the 2008 version of the Internet different from the 1969 version? The Internet is not something distant and foreign only available to IT specialists in the US Defence Force. We perceive it as a tool to interact with our environment, to keep abreast with the world in its totality, even if it means chatting on-line with a friend in the US about the Academy Award won by Charlize Theron, a South African born actress, using your own Internet connection at home in South Africa. The digital revolution unlike previous revolutions such as the industrial revolution is not controlled by external groups, political parties or governments but by the individual.

2.2 Cyber Psychology

In the physical world the words of Satre holds true 'I am my body to the extent I am', therefore one body can host one identity. The opposite is true in the virtual world where information is more important than physical appearances therefore the possibility exists to create multiple identities. During chat-room interaction the individual has the freedom to develop a 'pseudo-personality' whereby he/she can adopt any age, gender, name etc. With this pseudo-personality the individual has the advantage of interacting with everybody without showing the 'real me'. This discussion will deal with the various issues around the development of the pseudo-personality such as virtual interaction characteristics that promotes pseudo-personality development. The developmental process itself will then be explained in more detail by applying theoretical assumptions from the person centred approach of Carl Rogers. This section will be concluded with a discussion around the relationship between the real world and cyberspace with the pseudo-personality as vehicle to commute between the two.

2.2.1 Virtual Interaction Characteristics that Promote Pseudopersonality Development

We live in a world where sight is one of the most important senses used to evaluate the world as well as the people in it. Almost everybody has been exposed to Hollywood and the famous notion of acting out different roles and playing different lives also influenced virtual interaction. Although transvestites can be perceived with negativity in the physical world, on-line gender swapping is rife. The virtual environment also allows space for those that are different from mainstream, those people with the weird and sometimes wonderful ideas to meet people 'like me', thus combating loneliness. But let's start the discussion by keeping the next the well-known saying in mind: 'Beauty is in the eye of the beholder'...

a.   Invisible Appearances can be Deceiving (WYSIWIS - What you see is what I Say)

The 'WYSIWIS'-principle was first quoted by Gackenbach in 1998. During face-to-face interactions people infer qualities for their identities based on physical characteristics such as gender, race and clothing. During online interactions, physical appearances are invisible and cannot be assessed, at least initially, therefore friendships are built on different grounds such as similarity, values and interests. In the computer-mediated world the self becomes fluid and multiple, made and transformed by language and you are recognised through your textual behaviour by means of:

- Your signature
- Your nickname (According to Wallace [1999] 45% of nicknames are related to the individuals themselves in some way)
- How you conduct yourself in the chat-room
- The type of contributions to the conversations and
- Your unique language use (Wood and Smith 2001).

By means of their textual behaviour people present the idealised version of the self (i.e., those attributes the individual would like to possess) during online interactions. The reflection projected back is then one with good looks, irrespective of actually being overweight or the extrovert, but the real person behind the PC is actually the shy girl without the courage to talk to other girls on the playground. To summarise: 'We live in each other's brains as voices, images, words on screens' (Rheingold in Turkle 1995: 235).

b.   Morfing – Gender Swapping

Gender as demographic variable is one of the primary means by which we classify ourselves, a principal cultural box according to which we organise and make sense of our experiences. In most cultures the self is treated as being located in one body with a male/female dichotomy. As stated earlier, during multiple chat-room interaction the person can have a body and be disembodied. Although the embodiment is unknown in cyberspace, Smith and Kollock (2001) stated that we will anticipate a body talking back to us that is either male or female. Wallace (1999) describes the same process but links this spontaneous gender evaluation to priming. When priming occurs, some social categories are more accessible than others and we tend to use the one closest to the surface to form impressions about newcomers. The male/female label is one of the primary categories by which newcomers will be evaluated. This is evident early in chat-room conversations when the question is asked: 'M or F?' (Male of Female), which was conveniently given the term 'MORFing'.

The 'MORFing' phase provides no insurance that the individual will portray his/her embodied gender. On the cyberspace stage, role play is rife and gender swapping does take place as is evident in the example recorded by Winter and Huff (1996: 52): 'I have always used my first name as my login id, and I don't want to change now, but if I had to do it over again, I would probably use my last name. I have discovered the disadvantages of being identified as a woman by my login id'. Smith and Kollock (2001) also found that women tend to choose gender-neutral names to avoid online harassment. Males tend to choose female personalities because it is perceived that females get more attention; receive more help as well as sexual advances especially if they are 'newbies'. Interestingly enough Wood and Smith (2001) found that those individuals that swap genders online tend to portray rigid stereotypical roles and are caught it the act more than often. This process was described as 'hyper-gendering' by Smith and Kollock in 2001.

c.   Reading Race Online

Erving Goffman, developer of the Impression Management Theory stated that some impressions and expressions are 'given' i.e. those deliberate actions to create a specific impression.   Then there are the impressions and expressions 'given off' i.e. those impressions that are more subtle and harder to control.   According to Goffman race is also a sign 'given off' and during online interactions race can be determined by focusing on ethnicity.   Ethnicity is based on cultural markers of membership to a specific group for example language or religion.   Race specific frames develop for example South African blacks tend to use the words 'sisters', 'bro' and 'comrades' quite often.   The racial identity can also become known to the audience by means of descriptions of the person's heritage, home-town, parents etc. (Smith and Kollock 2001). Although the pseudo-personality is almost a given, at least initially, during online communication, the 'real-me' do leave fingerprints behind that can be identified at a later stage.

d.   Marginalised Social Identities

Chat-rooms can focus on specialised topics and interests and many of them are not concerned with the mainstream and non-stigmatised issues. These chat-rooms play important roles in the lives of those individuals who possess concealable stigmatised identities.   It can be very difficult to find other similar identities in real life because of the potentially embarrassing nature of the identity and fear of the possible consequences of disclosure.   Anonymous online chatting provides the protective cloak needed for these individuals to admit to having marginalised or non-mainstream proclivities that must be kept a secret in real life (McKenna, Katelyn and Bargh 2000). For people who felt that their identity carried a stigma with it in real life, the Internet provided a safe environment by which they could find 'people-like-me' who help them build significant relationships with others. (Amichai-Hamburger, Wainapel and Fox 2002: 125).

e.   Loneliness and Self-disclosure

Another aspect that is interrelated with the levels of self-disclosure is the feeling of loneliness.   For the purpose of this chapter we define loneliness as the 'self-perceived state that a person's network of relationships is either smaller or less satisfying than desired' (Leung 2002:

242).   Many times lonely individuals are unwilling to enter into interpersonal situations that involve risk of being rejected, embarrassed or disappointment.   During chat-room interactions those risks are few because of the face-less encounters as is evident from the following verbatim response.   'I know that I sometimes use the net to combat loneliness and isolation, but I am usually content to read messages it makes me feel involved, and in fact through the notices, I get involved with things that otherwise would have passed me by.   It's like radio, only interactive' (Winter and Huff 1996: 30).

2.2.2   The Missing Link - The Journey from the 'Cyber-me' towards the 'Real-me'

Modern academics debate the differences and importance of the 'virtual' versus the 'real'.   In some cases the virtual reality is an idealised utopia, superior to the physical reality.   This study will show that the characteristics of virtual reality of the Internet i.e. anonymity and individual control over revealing information are green lights that attract those individuals who find it difficult to express the 'real-me' in face-to-face encounters.   It will also show the importance of the link between on- and offline living.   Cyberspace should not be seen as an alternative space something distanced from the physical world.   Cyberspace is an extension of our physical world, a world mediated by means of our computers, an electronic mirror reflecting the world we currently inhabit.   It will always mirror our real selves, our real lives.   Pseudo-personalities are temporary vehicles constructed by people for the journey of actualisation.

2.2.3   Web Spaces – The Mediators Between On- and Offline Living

The 21st century can almost be defined as the fast paced information age where if you blink you miss the opportunity to conclude your own memories.   Living has become moments of information that is revealed for mere seconds and then attention moves on to the next source.   Humankind adapts to their ever-changing environment by establishing yet another dimension to living; existing and interacting as different bodies in multiple web spaces.   These spaces are not screen-based imitations of printed information but are unique in that their content cannot be exactly expressed in other forms, because the 'medium is the message, and the web is a medium of interactivity.' (McKelvey 1998: 6).   For the purpose of this chapter Internet interactivity is defined as a response albeit from another person or the system. Visitors are connected through a carefully designed interface that adjusts to their needs or specifications whenever they are communicating online. Communication mediums as we knew it in the offline world has changed. Web spaces were created to:

- Provide access to a global audience thereby gaining information about the world and its different cultures through new ways of thinking about evolution, relationships, sexuality, politics and identity.
- Allow users to interact with information and the provider and so information flows both ways;
- Act as places where 'virtual' communities meet to share interests, ideas, norms and values and interests (McKelvey 1998).

Web-spaces creatively mediate communication where individuals, regardless of past or present expectations get together to express how they experience the changing world on a daily basis. More specifically in a chat-room being human are emphasised by telling the world, or and individual what we feel and what makes sense, whether that entails replaying personal events, asking questions or full out swearing. In web spaces there are no need for all our limbs or senses to work, or exist for that matter, we can simply connect to the Internet and trust the visual artists of cyberspace tot take us closer to where we can be part of the human race exactly the way we where intended. Visual artists of cyberspace create virtual spaces wherein physical bodies can interact without meeting each other physically.

3. Research Methodgy

'Don't waste your time trying to learn the tricks of trade, learn the trade' (Author unknown).

As African researchers educated within the Western paradigm we have found the value of first living in Africa and then applying Westernized research methods to study the African way of living. Results are based on:
- Content analyses of online behaviour
- Evaluations of how .co.za-websites' visual cues influence behaviour.

3.1 Units of Analysis for Internet Research

Many times researchers find it difficult to connect the macro level of the analysis of social systems with the micro level of the individual person and the transaction processes between the levels. This study aimed to address this problem by using the **three-level (domain) model** of social context (Mantovani in Guiseppi and Galimberti 2001: 22). Links between the levels can be studied in both directions starting from the use of the computer or from the social context influencing interaction. Individual computer interaction influence interaction in everyday situations leading to cultural changes on a macro level. In the opposite direction, cyberspace culture supplies tools needed to interpret situations correctly; situations generate aims that determine local interaction with other actors via our computers.

Level 1: Interaction (Micro level of individual interaction)
Carl Roger's ideas and thoughts around creativity described behaviour, goals and personal motivations of an individual to form relationships with other individuals via the computer.   Furthermore this level forms the basis or starting point of the development of cyberspace culture.

Level 2: Situation (Intermediary i.e. Group level)
Cyber-groups (chat-room groups) develop within the cyberspace environment with specific constraints and potentialities that differ from the traditional face-to-face encounters.   Specific roles, functions and behaviour such as pseudo-personality development describe this unique electronic situation people encounter on a daily basis.

Level 3: Context (Macro cultural level)
On the broader contextual level the cyberspace culture is evident, that provides meanings that are internalized, hence showing the dialectical relation between culture and social memory. Specific social rules develop and are brought forward to the situational level and are provided to 'newbies' as social rules guiding behaviour in cyberspace. This culture then influences other institutions within broader society such as the globalisation of economies and social relationships across continents. This project will emphasise the continuing mediating process between technological innovation and human social change, implying that the cyberspace culture is indeed both a social as well as a cognitive concept, making it an ever changing prerequisite for communication in the 21st century.

3.2 Population and Sample

The formula used to determine a sample size of 384 is:

$$n = \frac{(1.96)^2 * 0.50\,(1 - 0.50)}{(0.05)^2} = 384$$

Currently the population is infinite, since it is unknown how many South Africans visit Internet chat-rooms. The following assumptions apply:

- $z = 1.96$ for the 5% significance level or 95% confidence level. This is the number of standard deviation units in the 'normal distribution' that will produce the desired level of confidence (significance).
- $p = 0.50$. We have no idea beforehand of what percentage of the population has the characteristic of interest, and then we must be conservative and take it as 50%. This is the estimated proportion of the population who possesses the characteristic of interest i.e. Internet chat-room participation.
- $E = 0.05$. This value determines that the sample result should not differ from the population figure by more than 5%. It is the allowable error. It states the expected deviation of the sample result (survey result) for the probable population figure, under the specified significance (confidence) level.

In summary, with 95% level of confidence it can be expected that the sample statistics (results) will not differ from the population statistics (actual values) by more than 5% if the sample is 384 and consists of randomly selected respondents from the infinite population.

3.3 Realised Sample Characteristics

The following table will provide a list and number of the different chat-rooms that were included in this study:

Table 1:   Sample realisation figures

Chat-room source	Cell sizes	%
Debating chamber on the news24 website	84	22
Mail & Guardian chat-room	83	22
Chat-rooms hosted by BBC	64	16
Carte Blanche chat-room	41	11
Sê-jou-Sê chat-room on the news24 website	41	11
Yfm Radio station's chat-room	39	10
Metro FM Radio station's chat-room	32	8
TOTAL	**384**	**100**

4. Research Results

This section will aim to present the results from the micro to the macro level. Furthermore it will be summarised with a focus on the cyberspace culture as it is experienced by the .co.za-tribe highlighting:

- **The tribal members:**
Quantitative results will provide a demographic profile of the SA Internet user.
- **Their voice i.e. the 'co.za dictionary':**
In South African chat rooms unique words/phrases provide the ethnicity created and owned by the .co.za-tribe.
- **Their lifestyles in online living spaces:**
Specific .co.za-websites that hosts the chat-rooms will be discussed by focussing on the different groups they hosts by their ability to act as communication mediators between on and offline cultural living.
- **Norms and values of the .co.za-culture:**
On a macro level, norms and values that characterise the cyberspace culture will be discussed illustrating how dialogue (i.e. chat-rooms) creates a social link between cultures across the globe to form the digital society.

4.1 The Tribal Members

On average 1 in every 15 South Africans accessed the Internet by the end of 2001 according to a report from World Wide Worx. This compares to 1 in every 2 users in first world countries such as the USA, Canada, Singapore, South Korea and Hong Kong. The August 2001 figures stated that the number of online Internet users worldwide was approximately 513.41 million. South African figures in ascending order were:

Table 2:   South African Internet penetration figure growth from 1997 – 2004

DATE	NUMBER	% OF POPULATION	SOURCE
2004	3,500,000	7.6	*OPA SA
2003	3,280,000	7.0	Goldstuck Report
2002	3,100,000	7.0	Goldstuck Report
2001	2,025,000	5.0	University of Pretoria
July 2001	1,500,000	3.7	Nielsen//NetRatings
December 2000	2,400,000	5.5	ITU
May 2000	1,820,000	4.2	Media Africa
August 1999	1,622,000	3.7	Media Africa
December 1998	1,266,000	2.9	Media Africa
November 1998	1,040,000	2.4	Media Africa
February 1998	800,000	1.7	South Africa Ondine
January 1998	600,000	1.5	SANGONet
February 1997	700,000	1.6	South Africa Ondine

*Source: Online Publishers' Association (South Africa) / **Source: (*http://www.nua.com*).
***Please note that these figures include Internet access in general i.e. at home, office or elsewhere. If an individual has access in more than one way he/she is only counted once.

4.2 Their Voice i.e.: The 'co.za Dictionary'

Our need for creativity in order to grow and adapt to different circumstances has been discussed already and this section will deal with the creative way in which people express emotions in cyberspace as well as the new words that were created within these chat-rooms.

'Humankind is divided by ideas, but find common ground in what they feel' (Author unknown)

Many social psychologists argued that written language is limited in a sense that no non-verbal contact is possible and the sharing of emotions is only possible to a certain extent. Although individuals differ in opinions and perceptions during online debates they share emotions in chat-rooms by means of 'smiley faces' or emoticons. Emoticons are symbols or abbreviations used to describe emotions to bring back the feeling to the conversations. The Metro FM and Mail & Guardian chat-rooms were examples of excellent aesthetic appeal to their users. People had a choice of various **smiley faces** to express emotion e.g.:

- Laugh out loud

- Angry

- Aggressive mocking

- (Get a
  life...SUCKERS!)
- Scary

- Naughty

- Embarrassed

- Insecure

- Unsure,
  confused

Various **punctuation marks** are combined to illustrate specific emotions:

:-D	-	Tongue in the cheek smiley
;o))	-	Surprised
:P	-	Tongue out smiley
!	-	Shout together with capital letters
;-)	-	Winking and smiling
)-:	-	Sad
....	-	Symbolising silence to either show respect or sarcasm
^^!^^	-	Devilish remark
ozzZZ	-	Snoring and bored

Complimentary to the above unique words developed spontaneously in chat-rooms conveniently summarised in the **.co.za-dictionary**:

- **General rules of thumb**

a.   In .co.za chat-rooms, people tend to mix English, Afrikaans and another African language. When a heated debate took place many reverted back to their vernacular specifically creating distance between them and the other group member(s).

b.   As in other chat-rooms globally the traditional grammar, punctuation and syntax rules didn't apply. Spelling, tense and grammar errors were common.

c.   Rap-battles occurred among young, black South Africans

d.   In the majority of cases people avoided long postings.

- **.co.za-words/phrases**

The next table includes unique words/phrases that people use in .co.za chat-rooms that are also referred to as 'slanguage'.

Table 3: List of .co.za-words/phrases

English translation	.co.za-word/phrase
Bicycle rage (similar to road rage - only directed towards cyclists)	Fietsrage
Brother	Bra
Cape Town	CT
Did everybody go home?	Effrrry1 gonne houme
Emigrants	SA expats
Enough	E-nuf
Favourite	Fav
Girls	Cherries
Greeting	'Ta' = Greet / 'Cya laer??' = See you later / Going for 🍵 and a ➡️ Back in a bit / Hola / Howat in the hooood!!! / Yo ... yo ... yo
Johannesburg	Jozi
Nothing more to say	Nm
Number one	Nambawan
Online search	'A quick **Google** on ostrich diet reveals....'
Poor ability to rap	'You rhymes be weak like a decaff black tea...'
Relax	I, am in need of a **chill** 💊 got a heck of a week coming up...
Sister	Sistah
South Africa after 1994	Transformania
Weed	Dagga joint / Holly weed
Zimbabwe	ZIM

4.3 Their Lifestyles in Online Living Spaces

These lifestyles will be discussed from the individual level i.e. the pseudo-personality at work, followed by a discussion of the micro-groups that develop spontaneously during interaction in these online living spaces.

4.3.1   Individual Pseudo-personalities Active in South African Chat-rooms

The discussion around pseudo-personalities will be structured as follows:
* A description about the use of pseudonyms online
* The MAMA-process (Moratorium - Achievement - Moratorium - Achievement)
* MORFING -Gender swapping
* Creating fool proof pseudo-personalities

a.   The use of pseudonyms online
The majority of the chat-room participants made use of pseudonyms and some websites provided original content or site-specific icons that participants used to describe their pseudo-personalities in more detail. In Yfm chat-rooms they were able to use colours should they need it to enhance the description. Some examples of popular Yfm icons:

By analysing the pseudonyms they chose, interesting conclusions could be drawn about the real, offline personality, thus making the pseudo-personality an electronic reflection of the real person behind the computer. Herewith some examples:
For many black South Africans it were important to emphasize their African heritage and they deliberately developed race specific frames that act as cultural markers of membership to a specific group. Some use their African names e.g. *Thabo*, and others created African pseudonyms e.g. *'T-girl'*, *'Ghettolova'*, *'Qabakazi'* or *'Tsotsi_gurl'*.
Sexuality was identified as one of the major themes across chat-rooms. This was also portrayed in the specific pseudonym choices since online sexual flirting was a popular pastime across all chat-rooms motivating a focus on characteristics that enhance physical and sexual desirability. The majority of females emphasised their sexuality during online interactions e.g.:
Deliciouszz; _TeeZer_; Fatfree; Spiced chocolate (Yfm chat-room); *TellAll* (Mail & Guardian)

 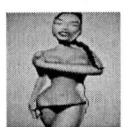

Some might argue that the only way to get noticed is to project sexual desirability. Many females across the different chat-rooms used their sexuality to introduce themselves to the group, thereby ensuring feedback / attention from males in the group.

Their black male counterparts also followed suit with pseudonyms such  as *'Nevasloppy', 'Testosterone', 'Stunner'* or *'Penetration'*. In Yfm groups, the males favoured the bodybuilding icon to illustrate male strength and sexual desirability. White males preferred external cues e.g. the car they drive for the same purpose and create pseudonyms e.g. *'GTI'* (Volkswagen Golf drivers).

They also used pictures to illustrate the barriers in offline living that limit sexual exploration and growth such as:

Many times lonely individuals are unwilling to enter into interpersonal situations that involve risks of being rejected, embarrassed or disappointment. During chat-room interactions those risks are few because of the face-less encounters and they find it easier to express their emotions and perceptions. You'll notice them by their pseudonyms:

alien8;   *old man*

Many individuals are desperately lonely and they create pseudonyms such as 'DONOTDELETE', 'Don't Speak' or 'Bliss' illustrating their willingness to please and do almost anything for acceptance, even if it means adopting an attitude of 'Ignorance is bliss' overlooking individual differences to alleviate their personal feelings of loneliness.

South African emigrants used .co.za chat-rooms especially the Mail and Guardian forums to keep in touch with their country of origin. Many of these emigrants feel alienated and alone in their new country and will use pseudonyms such as *'Goner'* or *'SAVANCOUVER'* to stress their need for social interaction with people 'like-me'.

You discover yourself by reflection from the people around you. With the freedom of identity construction, it might be that the individual has a need to be accepted and will turn into the person he/she thinks will get the attention needed. That is why many individuals chose pseudonyms such as:

'Little_miss_sunshine; JollyRoger; Funny man=online,
offline=Journeyman; Ed the Red

They project an extrovert, fun-loving, positive personality to conceal offline feelings of low self esteem and a lack of confidence represented in both pictures of the sad clown and 'Mr Bean' the lonely comedian.

Some individuals have a need to be admired by their online peers and they will introduce themselves as being on a higher level than the rest of the group e.g. 'Prophet', *'Gandalf'* or *'God of da Net'*. These individuals can be described as your typical 'wallflower' in his/her offline social group, the introverts that lack self confidence, usually the victims for bullies during online power struggles.

a.   The MAMA-process (Moratorium - Achievement - Moratorium - Achievement)

The digital identity is described and a 'licence to grow'. Within the ever-changing virtual world the individual is given a chance to discover parts of the self that he/she would not have found in everyday life. By adapting the MAMA-process the moratorium phase is a phase of self-doubt about who we are and we experience this every time we experiment with different roles on the Internet, those people or personalities we are online, those that we never got around 'being' in real life. Therefore chat-room participants choose pseudonyms like:

Questioning, = online, Addict = offline

Coupled with a with a picture of young girl (An adult female presenting herself as an adolescent girl in the phase of questioning things)

or

Headroom = online, Journeyman = offline

Coupled with the following quote:
'*I can't wait for a second childhood; I blinked and missed it first time round*'

b.   MORFING -Gender swapping

In the majority of .co.za chat-rooms, gender is one of the primary categories used to evaluate and classify group members. If gender neutral names were chosen, other group members made mistakes resulting in online flame wars e.g.:

> "You mean he's not Xhosa? / Sorry to disappoint you. She is Xhosa. / Stuart? Xhosa? and female? right..." (Debating chamber on news24.co.za)

c.   Creating foolproof pseudo-personalities

Goffman's theory on Impression management stated that some impressions and expressions are 'given' i.e. those deliberate actions to create a specific impression.   Then there are the impressions and expressions 'given off' i.e. those impressions that are more subtle and harder to control. The latter becomes evident in the specific writing styles and language the person use. It is obvious that 'Cypha Cat's' new personality was only skin deep:

> "CYPHA CAT WHY DID U CHANGE UR ICON AND NAME?" (Yfm chat-room)

During a discussion in a Mail & Guardian group the topic of 'Trolls' came up. 'Trolls' are those individuals taking on other existing personalities without being sensitive to the impressions and expressions personalities 'give off' in chat-rooms e.g.:

"He is probably back trolling under another nick. Why do you think he's one of the trolls HJ? Trolls' usually give themselves away big time by their typing styles anyway. (One from the old forum always spelt one or two words very very peculiarly, and when on this one, they do the same thing)."

4.3.2 Characteristics of (micro) Chat-room Groups

The previous section dealt with the different pseudo-personalities that were active in .co.za-chat rooms. Although each personality is unique there are similarities between them whether it be in terms of their interests, motives, online behaviour, needs etc. The following section will aim to cluster these groups together in terms of the specific chat-rooms. Each chat-room catered for one or more specific groups of people.

a.   Yfm and Metro FM chat-rooms
The qualitative analysis of chat-room behaviour on the Yfm and Metro FM websites yielded interesting results on this specific group of 'Black young yuppies'. Compared to the majority of uneducated, poor black adolescents living in squatter camps, this group is seen as the minority, not necessarily accepted socially because of their higher socio-economic status.

On entering the website the station is marketed with original content and slogans like: *'Listen to the Youth Speak'*. Furthermore The forum targets young adults (25 - 34 year olds), and aspirational youths (16 - 24 students & executives). The aim is to provide a tool for networking and interactivity while reflecting Metro FM's brand values (Yfm chat-room host).

This group of young, upmarket black South Africans portrayed needs to create an unique identity and a shared character of being proud of their African heritage, culture and customs. Cyber communication establishes and allows individuals to be part of new communities. It strengthens group solidarity and internal cohesiveness and they described themselves as part of the *'Y-family'* or the *'yworld.co.za'*. It seems that they knew each other offline as well, since they exchanged e-mail addresses and went to parties that were advertised in the chat-room. Sexuality is almost part of every discussion. It was evident that this group is disappointed with their existing political leaders and are openly, aggressively criticizing them. One might argue that the black background on both websites facilitated their outspoken and overtly aggressive attitudes towards authority re-enacting gangster-type behaviour and attitudes:

"Who cares? Its a free country u are free to do wotever u like as long as u
ar comfortable, others might not use vulgar for certian reasons, others uses
it for fun or they are used to shit out mouth, it wont hurt mos so what the
hell... FREE UR MIND AND KEEP UP THE POWER" (Yfm chat-room)

Very important was the 'cool image' that was projected at all times.
Slogans on the website like *'Ya all feeling this?'* promoted this attitude
and image, They also use 'rap battles' during online interaction, copying
American rap artists as role models. With their existing identity crisis the
title of 'EMENM-wannabee's seems fitting. He is a white male who is
famous for his Rap-music wherein he questions and addresses society's
problems e.g. racial discrimination etc. It seems that these adolescents
didn't have a need to meet those different from themselves. They only
used their 'rap battles' to comment on each other's 'coolness' and they
didn't really discuss issues such as HIV/AIDS, racial segregation etc.

"geeezzz people doesn't anybody wanna discuss the actual topics posted
here? This is a great medium for you all black privileged youth with
access to the Internet and you're not using to change anything. It's a damn
shame!" (Yfm chat-room)

b.   Sê-jou-Sê chat-room on news24.co.za
Advertised as an Afrikaans group, the traditional, conservative, right-
wing Afrikaner interacted here and many of the 'co.za-dictionary' words
spontaneously developed in this chat-room. Chat-room moderators did not
play active roles during conversations and racist remarks were the norm,
as well as underlying sexual tensions and conflicts especially concerning
the role of women. Many 'sex traders' from both gender groups spent
time here looking for partners open and available for cyber-sex. As the
black young yuppies, this group was also not interested in meeting others
'different from me'. They also knew each other offline and communicate
on a daily basis. With limited advertising and no specific topics for
discussion, not much has been done to attract 'newbies' to the site. A
newcomer could perceive the conversation as senseless and only directed
towards the in-group i.e. a group of e-friends knowing each other
personally.

c.   Mail & Guardian chat-room
In   this   chat-room   lonely   individuals   (wallflowers)   seeked
companionship and friendship. They were able to get to know their e-
companions quite well since the chat-rooms were structured by starting
with the topics under discussion, followed by the number of viewers,

threads and posts, ending with an indication of who posted something last and the time of the posting. In addition additional information was also provided such as the number of registered users and a personal welcome note for 'newbies'. The 'Shout Box' provided an additional interactive outlet for daily frustrations and was updated with new posts continuously. This chat-room also created a safe space for marginalised social groups that consisted of society's misfits.

Based on the informative nature of being an on-line newspaper, another group of people attracted to the chat-room were the 'informed tax payers', the well informed educated individuals, of higher income groups interested in debating issues in the newspaper, on-line in the chat-rooms and during debates hosted by the SA FM radio station.

A dominating trend (visible across all the chat-rooms) was the emerging the '.co.za-voter'. They openly criticised political figures and visibly reacted to the past where voters were described 'passive followers'. Finally you'll also meet the 'young at heart adventure seekers'. They saw the Internet as an opportunity / vehicle to explore unresolved issues e.g. sexuality, racial issues. It is a safe space where the process of 'growing into yourself' can continue.

d.   BBC chat-room

The BBC Prime television channel on DSTV is one of the popular television channels enjoyed by the high income groups resulting in website and chat-room usage as well since the content on both media types complimented each other. The 'leisure time expert' found another meeting space outside .co.za-borders where he/she could meet people with similar interests/hobbies. The site was structured by highlighting specific topics for e.g. culture, lifestyle, news, sport, teens etc. and should you get lost there was a message board sitemap for navigation. Slogans like: *'Talk your thoughts, your views, your space'* encouraged marginalised social groups to meet here as is evident in the next quote by a scientist:

"Science is not a democratic process - which is why great scientists are so often social outsiders".

Here the '.co.za-voter' met their .uk.za-partners since they also openly criticised political figures and their decisions. Slogans like *'iCan: Change the world around you'* facilitated discussions between action groups and as their Carte Blanche counterparts they focussed on specific self help groups for e.g. drug addicts where group members supported each other, provided information and advice, and a basic level of online group therapy.

e.   Carte Blanche chat-room

This chat-room complimented the M-NET programme each Sunday night and attracted the 'informed tax payers' with slogans like: Viewers do the viewing, through the Internet chats viewers are able to put their own questions to key players featured in the programme. Contrasting to the multi-racial Mail & Guardian chat-rooms these participants were mainly white. The emerging '.co.za-voters' were also very active here applauding celebrities and political figures that 'chatted' to the public on a horizontal, equal level.

Another spontaneous trend visible in this chat-room was the 'self help group' e.g. drug addicts, HIV/AIDS victims etc. Since Carte Blanche reported on highly sensitive topics at times, it provided a space for people experiencing similar problems to communicate e.g. parents experiencing parenting problems. This trend was identified during the quantitative analysis where we saw an increase on Internet usage among new, mature and single parents. They used the forum as a space to meet others 'like me' e.g. men dealing with the stereotype of being 'uninvolved fathers', white men between 40-50 facing retrenchment, or fathers battling with disciplining their children.

Complimentary to the above, chat-room participants were surrounded with other interactive options such as libraries, story updates, an option to buy videos, online bookshops and online voting polls giving them as much information possible that can be accessed in the comfort of your home.

f.   Debating chamber on news24.co.za

This chat-room also attracted the '.co.za-voter'. Chat-room moderators did not play active roles during conversations and chat-room participants openly criticised political figures and were mainly white and black males, emphasizing their own individuality and group affiliation. Although these .co.za-voters also used the chamber to interact with people 'different from me' as is the case with the Mail & Guardian chat-room, flame wars were common. It can be attributed to rigid personality traits that did not leave any room to respect others although they differ in opinion. They often reverted back to stereotypes that resulted in personal attacks and insults. The online conflict situations were also not resolved, thus making the whole process a futile exercise for breaking barriers between heterogeneous groups. In many cases senseless chat were recorded due to the fact that the hosts did not provide any topics.

In 2005 additional chat-rooms were introduced and structured according to specific interest/hobbies e.g. Formula 1 enthusiast. Being one of the sites most visited due to the news coverage according to the latest AMPS 2004 figures, the 'leisure time expert' found another meeting space within the .co.za-borders and one could argue that the bbc.uk.za chat-rooms might loose their .co.za-tribe visitors.

4.4 Their Culture

On the (macro) contextual level, various norms and values developed during chat-room interaction. The following themes dominated and will be described in detail:
- Social values
- Sexuality
- Race
- Politics

a.   Social values
One main underlying social value of the cyberspace culture is respect for others. Whether the chat-room participants were involved in a heated political debate or discussing a drug addiction, the value of respect for others different from you was important although not everybody adhered to it. Flame wars usually erupted when heated debates turned into personal insults degrading and stigmatising the other person. Since chat-rooms give voice to marginalised social identities and minority groups, individuality, diversity and eccentricity should be respected rather than criticised. Cyberspace provides .co.za-tribal members with the liberty (with a social responsibility) to base decisions on personal meaning rather than peer group pressure.

> "…, in the words of Rodney King 'Can't we all just, get along?' could be possible if we all realized that we really are one race. That we really are one species, one people. What we lack is sympathy, empathy, understanding and compassion for anyone who is the least bit different from us. This includes, but is not limited to, 'race', religion and economic or social status." (BBC chat-room)

b.   Sexuality
Sexuality was already identified as one of the major themes in chat-room discussions and will be described in more detail in terms of the values connected to sexuality. Although cyberspace is synonymous to the theatre, a realm where one is invited to perform a variety of roles, the key is to reach the achievement phase in the 'MAMA-process' i.e. integrating the newly acquired knowledge of the self into one integrated, intact, responsible self. Online chat-rooms provide people with a safe space to discover parts of the sexual self that they would not have found in

everyday life. Due to their online anonymity, females perceived chat-rooms as a safe platform to openly *be* sexual, discuss sexuality without the fear of being labelled as 'cheap and easy'. The South African 'LoveLife' campaign aims to break down traditional communication barriers around sexuality and promotes open forums where young males and females are allowed to discuss their sexuality openly breaking down myths and stereotypes.

> "Most guys are clueless when it comes to expressing themselves in writing - they just DO project that cheesy confidence when you're in their presence. So, it must be doubly hard for them to get down about what they're all about" (Debating chamber on news24.co.za)

Coupled with the first value of respect many homosexual individuals see and use cyberspace as a platform to emphasise the importance of especting sexual diversity often lacking in the face-to-face world.

> "Why does society insist on categorising people into boxes? Straight, gay, bi, or whatever? Isn't it just possible that we are all moving in between each of the 3 main categories during our lives? Personally i feel that i do. Sometimes i feel gay, other times straight. Does that make me bisexual? Or does it mean that i am confused? Personally I couldn't care less because I am happy with whatever i feel. All i am trying to say is that when i read about other people who are worrying about what sexuality they are, i ask myself 'Does it really matter?' Who actually cares what sexuality we all are. It doesn't affect our daily lives, or the way we live those lives. We are all human after all! / I've never given myself a catergory, and as a result (or maybe not giving myself a catergory was a result of this, I don't know) I've never felt I fitted into any particular social group or scene if you like, and I've always been reluctant to define myself into a catergory. As I may have suggested, I don't really care about catergories anyway" (Mail & Guardian chat-room).

One dominating theme evident across the chat-rooms was the lack of sexual maturity. Unresolved conflicts in terms of sexual relationships were also revealed during online interaction. Some individuals have never learnt effective and respectful ways to deal with sexuality when members of the opposite gender are present. This was evident in their distasteful, vulgar approach to sexuality. Sexuality was also used to degrade and insult each other and/or political figures for e.g.:

> "That's what your mother tells your... hang on you dont know who you father is, so I guess that's what she used to tell her latest pomp every night... / I know who my daddy is... ...Wanna know who your is? Huh?

Who's your daddy boytjie? Thabo, that's who! Whether in London, or Orania, makes no difference. Don't you forget it baby. Now fly away tweety-bird.. Go advertise you racist ignorance to someone who takes you somewhat seriously. Seriously dude" (Debating chamber on news24.co.za)

"(Swearing) (political figure) Can't y'all, see that he's fake, Won't break you a crumb of the little bit that he makes And this is with whom you want to place your faith? Cause the (swearing) wear a coofie, it don't mean that he bright Cause you don't understand him, it don't mean that he nice It just means you don't understand all the (swearing) that he speak. (Swearing) (political figure). to u flipchik: I'm a legend, you should take a picture with me You should be happy to be in my presence, I should charge you a fee I'm Big Dog, listen (political figure) you a flea And the little phoney (political figure) is a garden to me What's the problem (political figure)? You not as hard as me." (Yfm chat-room)

Black females in the METRO FM and Yfm chat-rooms were more outspoken and open about their sexuality compared to the other women on different websites.   They used, almost flaunted their sexuality over shadowing other personality traits and characteristics e.g. intellectuality during debates. It might be that this group were highly influenced by the black female singers e.g. Beyoncé, 'JLO' (Jennifer Lopez) and Janet Jackson, all well known for sexual daring behaviour and music videos.

Online dating was also a popular past time for .co.za tribal members for various reasons:

- **Some enjoyed the humour attached to the process.**

  "It's the dating thing on News24.… But if you want a really good laugh, say that you'r a woman looking to meet men, and check out the pics and pick-up lines.   It's really ba-a-a-a-d!" (Debating chamber on the news24.co.za website)

- **Others especially women preferred online dating since it provides the freedom of an 'emergency exit' out of an uncomfortable situation with a stranger.**

  "You dont' need to put yourself out 'like that' you meet person, maybe you like or dont', you dump person and keep surfing till you find someone you're feel most compatible wtih, its a stress free and painless exercise, cuase when you're do on screen dumping, you dont' have to feel guilty or sorry about, minimise risks of being emotionally blackmailed and end up feeling guilty and carrying on in a relationship you dont` want to be in, cause of guilty feelings" (Debating chamber on the news24.co.za website)

- **The pseudo-personality fluid and multiple, transformed by language was also put to use for both gender groups.**

"Notice, though, how few women place their pics there, as opposed to the guys? with girls its not so much about the looks its the heart!! pliz keep that in mind"

"Noticed that...Hmmm. Men lie? Average body = beer boep(tummy)" (Debating chamber on the news24.co.za website)

- **Those engaging in online dating were aware of the risks involved.**

"But one`s radar must be in top condition when engaging in a Net love affair. Too many horror stories. Fraud. Heartbreak. Murder. Worse- Seeing a pic of a delectable dish of a male specimen, only to get a Quasimodo."

"You know one always knows that you`re taking a risk with these things whether on/offline, whatis to guarantee me that a guy I meeet in the pub is not a serial killer?" (Debating chamber on news24.co.za).

- **Despite these risks many took the online dating game seriously aiming to meet life partners and succeeding.**

"I was also sceptical when this girl contacted me from SA and told me about her online romance, but when she eventually married and I got to meet her beau, she is really happy, thats what counts, and this guy is very good to her kids, that for me is the main thing"

"One of my best friends met her hubby on a chat site.. they have been married a while now with kids too....so there is magic in all places I guess....." (Debating chamber on news24.co.za).

c. **Race**
With our segregated past it was no surprise that the racial issue came up as another dominating trend. People from different racial dominations still battle to integrate different belief systems and still revert back to degrading, stigmatising and insulting each other. However chat-rooms were used as public forums that stimulate interaction between heterogeneous groups since cyberspace is controlled by individuals and not political parties and governing bodies.

"How would this forum be different had apartheid never ended? 1) Do you think the government would have allowed uncensored access to the Internet? Would a forum where foreigners could participate be okay? 2) Do you think most people here would be defending apartheid or not? 3) Would there be discussions of immigration, sanction-busting, violence in the homelands, or would most people not know that there were any problems at all? 4) Would there be any blacks/coloureds/non-whites? Would the government have declared the 'net slegs vir blankes or discriminated de facto, not de jure? PLEASE give me your thoughts - this fascinates me.

Yes, things would be a lot different. In response to your questions, in order.... 1. The government would never have allowed uncensored access. I believe that foreigners may be allowed to post, but with restrictions, ie. posts vetted before being allowed to be 'shown' 2. Yes, most people. You see, so many many of us didn't really know what went on... I had a taste of it as a white person... threats against me for associating with black people, for defending them, for expressing my disgust.. it was horrible, but I can still never ever grasp the enormity of the cruelty inflicted against my people. It's overwhelming and so very very sad, and although I never harmed anyone with either word or deed, I do feel bad because I did not know enough/do enough, to protect the people I love so much. 3. Discussions perhaps, but limited to knowledge... there was a great deal of censorship and propaganda.4. I don't know if they had the capabilities to determine that, but if they did, they would probably have said.. net blankes... I, as a white, female adult, am so distressed at what went on, and still goes on. I've witnessed it often, and would want to wish it away. The nice thing though, now, is that if I do witness it, I can speak out without fear of getting arrested or beaten up. --------------------Despite all allegations to the contrary, a true and honest patriot. Go put that in your pipe and smoke it, detractors!" (Mail & Guardian chat-room)

d.   Politics

Complimentary to the racial issue was the political situation in the country. The .co.za-voter openly criticised political figures (national and international). Although very outspoken about various issues such as economics, education and health, political leaders were constantly being evaluated in terms of their actions on a personal and public level.

"I consider (political leader) to be the most dangerous man in the world' / (Political leader) is an excellent and clever campaigner, even if he is a very stupid and dangerous man, while (Political leader) seems to have undergone a personality bypass operation, but otherwise seems fairly sound" (BBC chat-room).

Politicians and their decisions were questioned continuously against the background of their respect for the man in the street. The .co.za-voters put politicians on the same level as the public acknowledging the mutual respect between them as humans. As discussed before, respect for individual integrity and honesty where you have to be true to your own beliefs and value system were very important to .co.za-voters. Politicians operating within the cyberspace culture must be willing to come closer to the people on the ground, narrowing the gap between existing authorities' societal structures and the voters. The .co.za-voter will respect a politician in stead of fearing his/her authority.

> "(Local political party) wants African people to join it but it does not respect their cultures and values" (Debating chamber on news24.co.za).

> "What IS wrong with us. The simple answer, I suppose, is fear. Regardless of our age, sex or race, we are scared of our leaders. We are scared of their guns and soldiers, their militia and power. In four years we have watched the price of a single loaf of bread go from ten to three thousand dollars and yet we are too scared to do anything about it". (Mail & Guardian chat-room).

5. Internet Research in the .co.za-tribe

South Africa is a very heterogeneous country with both first world and third world characteristics. Global companies often embark on global research projects in order to inform their business strategies and marketing plans. It is of great value if their selected methodology could be applied uniformly in all the markets being measured to ensure as much comparability between the markets as possible. These companies often find it hard to understand that the same is not always possible in South Africa, especially in terms of Internet research. Internet penetration is very low in South Africa, which limits the groups that could be targeted for this type of research. Internet access is also viewed as relatively expensive compared to the rest of the world. To elaborate on the applicability of on-line focus groups in the South African market, Richards (2005) interviewed 10 South African researchers and psychologists.

The chat-room as research tool, equivalent to an on-line focus group is a 'foreign' concept to local researchers. These researchers thus still prefer traditional face-to-face focus groups, even when targeting .co.za-tribal members. Few traditional researchers will see it as a necessity for innovative local research projects. Comments by cyberspace researchers operating in the local research market suggested that the '.com-research toolbox' is a nice-to-have tool but is expensive and difficult to implement. They also felt that it was an appropriate methodology in research aimed at a specific captive Internet audience or where international clients insisted on its use. The majority still felt that crucial information on non-verbal behaviour was lost whilst e-groups did not effectively project emotions and other non-verbal gestures. Authenticity and representivity of on-line samples are still being questioned due to the lack of appropriate measures and infrastructures to implement valid and reliable cross-checks.

The '.com-research toolbox' currently includes limited research tools to reflect true real life experiences of South Africans. Passive observation of chat-room behavior is one effective research tool that can be used by researchers. E-mail surveys generally have a very low response rate, and many times, on-line questionnaires are printed and sent back to the researcher via the traditional route such as fax machines.

Ethical Internet research practices remain a proverbial hot potato since the majority of the .co.za-tribal members still do not trust website owners with personal information such as account details and ID numbers. In fact, the debates around ethical human-electronic interface have been heard in many academic circles, but there still is a reluctance to test these theories in practise.

There is still a long way to go before the '.com-research toolbox' is accepted by the .co.za-tribe. However the possibility does exist to use the chat-room as vehicle for establishing on-line self-help groups since .co.za-tribal members feel comfortable under the protective cloak of anonymity provided by on-line chat-rooms for open discussion of personal and other problems, destructive behaviour as well as dependencies. In summary the on-line chat-room will remain an important medium where minority groups in society can voice opinions lost in the voice of society as a whole as expressed in other forms of mass media.

Conclusion

In conclusion although a small portion of South Africans use the Internet as another complimentary interaction medium it will continue to play a major role in our lives. This paper aimed to illustrate how technology specifically the Internet and cyberspace influence the South African way of living. We see ourselves as part of the global cyberspace culture, creatively changing our environment that is influenced by digital technology in the globalizing world. The .co.za-dictionary showed the creative way in which .co.za-emotion and memory is expressed and negotiated. The distinction between the virtual and the real, does not imply a privilege to either, but rather a connection between the two. The pseudo-personalities at play in the borderless world of cyberspace continuously reflect the issues, problems and struggles of everyday life in South Africa such as racial tension, political struggles and sexual interactions. These interactions reflect our 'memories in action' that create new social networks governed by cultural norms and values of respect, an openness to the unknown, looking towards to self (not governing bodies) for direction resulting in liberty with responsibility towards the common good ,the core of every society across the globe. In the words of Mark Shuttleworth, the first African that reached space:

> "The Net is still on course to become the fundamental platform for all communications... from your cellphone to your fridge, they will all talk TCP/IP. But the net makes competition brutal, so expect to work hard for those profits. The mistake was in thinking that something that removed all barriers to entry could also be a source of infinite profitability! I don't think we'll see any new Yahoo! or Es-Bay emerge, but we'll see great new ideas like Slashdot.org that find a place for themselves, driven by people who love what they are doing and do it better than anybody else as a result, using the net to reach their listeners at the lowest possible cost" (Carte Blanche chat-room).

References

Amichai-Hamburger, Y., Wainaple, G. & Fow, F. 2002. 'On the Internet No One Knows I'm an Introvert': Extroversion, Neuroticism and Internet Interaction. *CyberPsychology & Behaviour,* vol.5, no.2 p.125-128.

Conference on Creativity (1952: Granville, Ohio). 1952. *Proceedings.* Toward a theory of creativity. Edited by C.R. Rogers

Gackenbach, J. 1998. *Psychology and the Internet.* Canada: Acadamic Press

Guiseppe, R. & Galimberti, C. 2001. The Mind in the Web: Psychology in the Internet Age. *CyberPsychology & Behaviour,* vol.4, no.1 p.1-5.

Guiseppe, R. & Galimberti, C. 2001. *Towards CyberPsychology: Mind, Cognition and Society in the Internet Age.* Netherlands: IOS Press.

Leung, L. 2002. Loneliness, Self-Disclosure, and ICQ ('I seek You') use. *CyberPsychology & Behaviour,* vol.5, no.3 p.241-251.

McKenna, Y., Katelyn, Y.A. & Bargh, J.A. 2000. Plan 9 from cyberspace: The implications of the Internet for personality and social Psychology. *Personality & Social Psychology Review*, vol.4, no.1 p.57-76.

McKelvey, R, 1998. *Hypergraphics.* South Africa: RotoVision
Nua *http://nua.com.* (22 Aug.2002).
Pinkerton *http://pinkerton.emeraldinsight.com.* (11 Mar.2004).
Rogers, C.R. 1961. *On becoming a person: A therapist's view of psychotherapy.* London: Constable.
Sampson, S.E. 1998. Gathering customer feedback via the Internet: instruments and prospects. *Industrial Management & Data Systems,* vol. 98 no. 2 p. 71-82.
Smith, M.A. & Kollock P. 2001. *Communities in cyberspace.* London: Routledge
Turkle, S. 1995. *Life on the screen: Identity in the Age of the Internet.* London: Phoenix.
Wallace, P. 1999. *The Psychology of the Internet.* USA: Cambridge University Press
Winter, D. & Huff, C. 1996. Adapting the Internet: Comments from a women-only electronic forum. *American Sociologist,* vol.27, no.1 p.30-55.
Wood, A.F. & Smith, M.J. 2001. *Online Communication: Linking technology, Identity and Culture.* Mahwah: Lawrence Erlbaum Associates.

Additional Reading Material

Lauer, DA & Pentak, S, 1995, *Design Basics fourth edition.* Harcourt Brace College Publishers.
Rockport Publishers, Inc., 1997, *Interface Magazine.* V&W Publications.
Steinberg, S. 1995, *Communication Studies, an introduction.* Juta & Co. Ltd.
Skopee, D. 2003. *Digital Layout for the Internet and other Media.* AVA Publishing.

EX-PEAU-SITION:
NOMADIC BODIES AND NEW TECHNOLOGIES

TATIANA MAZALI

Abstract: The object of analysis is the body re-written by technologies: a frontier of communication that from the immateriality, flexibility and post-textuality of its supports and processes seems to lead to new forms of orality (Ong, 1982), at the base of new tribes (Rheingold, 1993) which have rediscovered the importance of the senses in the process of learning about the surrounding world and its tools/extensions (incorporated knowledge [Lévy, 2002]). The main field of my analysis is the creative act contained in techno-body performance art.

Keywords: Nomadic body, intimate technology, performance.

The Body: Virtualization and Nomadism

Body means departure and movement, action and experience. The body spreads out through the world by means and with the help of technology, and thus achieves the transition from organic to inorganic that happens in the bio-techno-logical Here and Now. From within the dimension — not merely superficial — that its skin 'embodies', the body accomplishes unprecedented exceptions and unfeasibilities in its aim to become the new stuff that could hold together the shattered pieces of a world divided between opposites: natural vs. cultural, human vs. mechanical, biological vs. digital, real vs. virtual.

By the very act of spreading out and exposing itself to the outer world, the body defines, places, positions itself: *peau-sition* in an *ex*, that is, position of the self outside, being exposed outside. Much more than merely spreading out a surface, displaying the body implies a positioning (re-definition, individuation, construction); it is an act of creation — the world comes to light through a knowing body — in the way described in the philosophy of Jean-Luc Nancy (1995), who coined the suggestive and at the same time enlightening expression *ex-peau-sition*: EX= exhibition, PEAU= body/skin, SITION= position/location/definition.

The body escapes fixity and endlessly repeats the *ex-scription* movement (passage from the inside to the outside, act of writing that generates the body by exposing it to the world). Consequently the body cannot be a thematic object, but remains pure happening, event — that is, exception.

Ex-peau-sition is the term that I will use here to define the body. It is my alternative choice in relation to the various concepts of body characteristic of western thought: from a bundle of organical phenomena to an organism in itself; form the minimal unit of meaning (body = sign) to the psychological seat of the soul; from the Christian body, torn between Heaven and Earth, to the secular and dionysic body; from the spied upon and punished body of Foucault, to the liberated body of the avant-garde art of the 60ies and 70ies.

Nowadays the body is confronted with problems that are due to the stream of new technologies; problems like *virtualization*, for example, or the *nomadism* issued from the dynamics of places/flows that characterize our contemporary world (Castells, 1996). By this confrontation, the body experiences a new centrality and is claiming back its former cognitive function: incarnated thought, corporal intelligence, tangible knowledge — the tamed body becomes the tamer.

The actions performed by techno-bodies, aside from acts of conscience, are real feats of writing. According to Jean-Luc Nancy, the act of writing, which is a technology, is a constituting part of the body in the movement of *ex-scription* (going outside). Writing is an act of significance, it does not correspond exclusively to the signs and images of the body, but places itself in the *touch* — first place of conjunction between body and world. Writing implies first of all touching, and only later signifying. The writing act, prior to being a writing *of* the body, is body itself, it is its essence.

Ex-peau-sition is thus the creative act that permeates all the art performances discussed in this essay. Each of these techno-bodies is exception and at the same time bifurcation of the schizophrenic condition of the body in our culture. Each one embodies the desire to recapture reality by means of an act of writing/scripting (the *ex-scription*) where the signs — which transform the body into memory — are often unerasable traces.

The traces left by the mutant bodies of the performances are political actions insofar as they put to evidence — through their exposition, which is movement, not fixity — the possibilities and the contradictions of the technological present. And these political actions are acts of initiation — distinction and simultaneously enrolment in a group:

The written bodies — incised, engraved, tattooed, scarred — are precious bodies, preserved, set aside like the codes of which they constitute the glorious enneagrams: but they are not the modern body, that body which we have tossed down in front of us, which comes to us naked, only naked, and *ex-scripted* in advance to any writing. The *ex-scription* of our body is the process through which we have to go in the first place. Its outside-inscription, its setting *out-of-text*, seen as the most *proper* movement of its text: the *very* text abandoned, left at the end of its limits. We are not even talking about a 'fall', there is no up and down anymore. The body is not dropped down; it stands on the limit, on the external edge, the extreme one, the one that nothing can hold. (Nancy, 1995: 13)

The ambivalence of the body — the body is *this* but also *that*: its ambivalence is *sense openness* (Galimberti, 1983) — is polysemy, a polysemy that becomes more evident today with mutations and technological interaction. The techno-bodies affect the relationship between organic/natural and technological/artificial/cultural, at the heart of which a reconceptualization of the body takes effect: it is not an unchangeable element of nature anymore, but a concept at the limit, a nest of fears old and new.

The construction of a boundary between nature and culture serves several ideological purposes; most notably, it guarantees a proper order of things and establishes a hierarchical relationship between culture and nature. At a basic level, this socially constructed hierarchy functions to reassure a technologically overstimulated imagination that culture/man will prevail in his encounters with nature. The role of the body in this boundary setting process is significant because it becomes the place where anxieties about 'the proper order of things' erupt and are eventually ideologically managed. Techno-bodies are healthy, enhanced and fully functional — more than real. New biotechnologies are promoted and rationalized as life-enhancing and even life-saving. Often obscured are the disciplining and surveillant consequences of new body technologies — in short, the bio-politics of technological formations. (Balsamo, 1995: 215)

Among the new dimensions that constrain to reconceptualize the body in the techno-logical modern-day, virtuality is one of the most complex.

Virtual space is never steady, it moves constantly due to the dialectics between image and language, image and model, presence and representation. Virtual space is a space of the language in the sense that every image calls to mind other images or models, bringing about a multidimensional hyper-image that blends different layers of meaning and perception. The insertion of the body and the senses in the field of virtual communication is entrusted to an alter ego, an electronic twin (*avatar*) which, by definition, is not equal to its physical original. But such an electronic mediation - such a radical separation between image of the body and body itself - is precisely what develops a high sensitivity in the organism: virtual reality activates and sensitizes the physical body — the body is handled in its perceptive dimension; the digital simulations aim at the senses of hearing, sight and touch, and their lack of precision makes the body actively participate through self-stimulation[1]. That is the paradox of the activation of the body in systems of virtual reality and in other forms of man-machine interaction: the highest level of abstraction in virtuality can only be attained by re-producing the body, by making it therefore a superior value.

Just as the representations of the body are not the body but can be corporal, virtual reality — generator of representations — is not the body but is indeed corporeal. Pierre Lévy (1995) insists on the fact that the process of virtualization of the body does not mean dis-incarnation, but rather reinvention, reincarnation, multiplication, vectorialization: a hetero-genesis of the human body. On one hand, with his essay *Qu'est-ce que le virtuel* Lévy has definitively put an end to the question of the opposition between virtual and real: he substitutes it with the opposition between virtual and actual. But, on the other hand, a rather more political debate remains open on the subject of the un-personalization induced by the virtual-reality technologies. The most common impression is that technology takes us apart from the real and personified world, be it by giving us new opportunities to get free from the strategies of political and economic control — historically based on the administration (property) of the material world (material goods, workers' bodies) —, or by simply offering us new illusions and un-realities.[2]

That is the theoretical context in which the virtual-reality systems evolve and develop, which shifts the problem from the question of democracy/control in the information to that of presence/control over one's own identity. These systems give the opportunity to feel the *other one*'s presence and to react to it by way of technologies that are mainly visual, sonic and tactile. The plunge into the artificial realm can be taken by means of various interfaces: some of the most common are *datagloves* and *head-mounted displays* (HMD)[3], to which *haptic interfaces* can be applied,

---

[1] Using the terms of McLuhan we could define VR as a 'cold' technology.

[2] There is a plethora of political writings on the virtualized body, and mainly in the field of *gender studies*. Among the most renowned authoresses: Donna Haraway (1991), Meredith Bricken (1991) and Allucquère Rosanne Stone (1996).

[3] Among the pioneers of the interfaces for virtual-reality systems Ivan Sutherland deserves special mention; his three-dimensional display of 1966 and the

that a computer may use to send physical feedbacks to the user in correspondence with his movements.

A substantial part of the research on virtual-reality (VR) systems is applied to clarify the concept of *presence* as a focal point for the construction of optimized and efficient interfaces, really appropriate for the man-machine interaction. I shall put forward in these lines the definition of *presence* established by Lombard and Ditton (1997): 'the perceptual illusion of nonmediation. The term 'perceptual' indicates that this phenomenon involves continuous (real time) responses of the human sensory, cognitive, and affective processing systems to objects and entities in a person's environment'. The impression of presence would then correspond to an immediate experience, where the intermediation with technological artifacts would be minimally perceived or totally invisible. Presence stands in correspondence with the transparency of the technologies[4].

In VR systems the sensory level stands in for the concept of presence. That is the goal, for instance, of the investigation on *haptic interfaces*, which can provide the users with physical feedback: 'Advanced telepresence and virtual environment systems are to provide a human operator (HO) with sensory feedback in various modalities of human perception. All existing developments comprise several subsystems, each generating stimuli related to one of the human senses: vision, audition, kinaesthesia, touch, smell, and taste' (Kammermeier et al., 2004: 1). Technological research points towards the polysensory nature of these interfaces as a way to overcome the typical limits of the virtual body: fragmentation and 'subtle-ization'.

It is the relationship between artificial space, structured by language or by a digital code, and physical body, in its dimension of sensory perception, that characterizes the VR systems - systems that immerse the user in a computer simulation - as well as the *ubiquitous computing* systems - technologies that immerse the computer in an existing environment. The latter have been described by Mark Weiser (1991) with the expression 'embodied virtuality': a virtuality integrated, let's say grafted, on an existing reality. In virtual reality, the senses of the user are reconverted to attain functionalities congruent with the simulation; in

---

subsequent stereoscopic HMD of 1968 aimed to provide the user with an image in perspective that would vary according to the latter's movement. For a historical overview on the evolution of virtual-reality systems, interfaces and technologies refer to Rheingold (1993).

[4] The issue of transparency of the technologies is related to the concepts of *Immediacy* and *Hypermediacy* put forward by Jay David Bolter and Grusin (1999).

ubiquitous computing, the senses act in the normal way but are given a wide range of possibilities by means of the computing system spread in the environment.

Adaptation and knowledge acquisition through experience are the two actions that the body confronted with virtual realities has to deal with: in the case of VR, the body must adapt its senses to the machine; in the case of *ubiquitous computing*, the body must understand what additional possibilities are offered to it by that encircling intelligent environment.

As Katherine Hayles underlines, the virtual body must undergo an experience, and that experience is not without consequences:

> Working with a VR simulation, the user learns to move her hand in stylized gestures that the computer can accommodate. In the process, changes take place in the neural configuration of the user's brain, some of which can be long-lasting. The computer molds the human even as the human builds the computer. (Hayles, 1996: 275)

The mutual restructuring of man and computer is brought about by means of a language predominantly retinic and tactile. In virtual reality what matters is mainly what can be seen or touched. The symbols don't manage to get structured in real texts, the image decomposition on tiny screens inside a helmet is not sufficient to produce an intelligible text, the scene is flattened to an instantaneous perceptive dimension. A virtual environment allows the user to see only the surface of the visual world, even if it generates nevertheless the impression of real presence.

If knowledge is *personified* — in the sense that we humans manage to know through our corporal situation (be it individual or social) but not through an abstract and unselfish thinking process —, then personified knowledge is the knowledge of a body in a particular situation. In virtual reality, as well as in other fields of electronic communication, personification prevails as the only possible way of knowing and the body becomes the place we watch the world from; 'In some ways, the body is a primordial display device, a kind of internal mental simulator' (Biocca, 1997).

The technologies of virtual reality put to disposal but also claim urgently a new form of sensitivity which produces that flexible corporal experience that Derrick de Kerckhove (1994) defines as *liquid*.

Flexibility and liquidity are the new potentials, and at the same time the new constraints, of the world transformed by the information technologies.

Virtualization makes a nomad of the body, a nomad in relationship to space-time coordinates that become less and less controllable, and also in relationship to theoretical positions that don't manage to give answers anymore to the sense of misplacement that derives from the aleatory character of many an encounter, experience or life event.

In its greek etymology, the word *nomás* defines the nomad as 'the person that wanders in search of new pastures', and we can trace in that term the link between nomadism and law (*nómos*) that I propose in this essay as a fundamental aspect of contemporary nomadism.

The law corresponds to the circumscription of a space[5], the nomad to the overcoming of that limitation. The law corresponds to the imposition of rules, the nomad to the triumph over these. I suggest in these lines that we shift the attention from a concept of nomadism strictly related to going astray in space and territory — nowadays formulated whether through the places/flows dialectics (Castells, 1996) or the places/non-places dialectics (Augé, 1996) — to a new concept of nomadism related to going astray from the norms and time-honored rules of our cultural and political tradition. This adjustment helps me emphasize the exceptional character of the nomad: exception versus rule, the nomad — understood as drifting away from the rules — becomes a paradigm of exception. The nomad is a political individual, he is potentially subversive and can put forward, by impersonating them, new settings that criticize our acceptance of rules and facts taken for granted.

Overcoming a limit and drifting away are two essential aspects of nomadism: an errancy from a space to another and from a time to another, a political errancy against the established order, and a physical and experimental errancy in quest of a freedom that could offer the individual creative opportunities. The nomad is also the person who is not 'armored by a goal' (Benn, 1992: 24). Therefore the nomad is not only the one that wanders in space, lacking a house, but also the one strolling through time,

---

[5] The space that Hannah Arendt describes as the space of politics, governed by laws violent by definition, since they impose a limited freedom: 'The law, in its greek sense, is whether agreement nor contract; in fact, as it does not originate from discussion nor from the confronting actions of persons, it does not really belong to the political sphere but is mainly the product of a legislator's thought [...] The law is the hill created and set up by a single man, on which the properly political space originates, where many evolve in liberty [...] The essential point is to draw the limits, not to tie up and unite [...] The law has come into being by fabrication, not by acting; the legislator is like the town planner or the architect, not like the statesman or the citizen. By generating the space of the politician, the law contains the factor of offence and violence typical of any fabrication' (Arendt, 1995: 87).

lacking a goal: he transgresses the sequential character of time and gives it a value according to the present instant.

Nowadays, with the passage from Modern to Post-modern, nomadism has lost its subversive sting, it is not a state of exception anymore, it has become a diffuse and pervasive condition. Our contemporary techno-social system has converted into a rule a type of existence which, in other historical periods, used to represent a real strategy to free/differentiate a person. The techno-media claim that nomadic errancy is the main way to access and crisscross one's own world, transforming the relation between the global and the local, the community and the individual. As Manuel Castells puts it, the information society is based on opposition and on a search for dialectical synthesis between the Net and the Self. On the other hand, by specifically describing contemporary nomadism in terms of a *glocal* nomadism, Michel Maffesoli emphasizes the present gap between globalizing institutions and localist movements, a gap that would explain those new socio-cultural neo-communitarian practices 'that set in motion different forms of nomadic errancy, which boost up with dynamism the deep-rooted immobility of the single individuals: motionless in front of their own screens and therefore present in every place' (Maffesoli, 2000: 15).

Maffesoli also underlines the rebellious and liberating aspect of the nomadic condition, especially when he declares that errancy — apart from being the foundation factor of any social grouping — perfectly expresses the plurality of the individual and the duplicity of existence, without failing to convey the idea of insurgence, be it violent or cautious, against the established order.

The question is to understand how the double dynamic of exile and reintegration that generates the nomadic status is working nowadays, and to try and understand in particular the nomadism of the techno-bodies. The socio-technical establishment has *domesticated* modernity and nowadays presents the techno-bodies as fashionable bodies, depriving them of their exceptionality and integrating them in familiar Media imageries. The subversive nomadism inscribed in the techno-bodies — the techno-body symbolizes the way biology overcomes its limits when confronted with technology, it symbolizes the exile from biology and the reintegration in bio-technology — has become today a sweetened and cheap version in media communications that tend to annihilate exceptionality by diluting it in overflowing image repetition.

Pierre Lévy sees contemporary nomadism in an even closer relationship with the technological context: for him, modern nomadism is the result of a mutation induced by the new media and is highly dependent on migration, as a new condition based on the permanent motion of territory. Present-day nomadism is mainly determined by the unrelenting and swift transformation of landscapes — scientific, technical, economic and mental landscapes —, the consequence of which is movement. A movement to which, on one hand, we have to adapt permanently but to which, on the other hand, we can never adapt since reality is not a stage previous to us but the constantly changing result of what we do. For Lévy, the construction of reality is a collective task, and this makes present-day nomadism also collective; it even becomes the representative image of the new technological mankind:

> The conquest of space aims at a radical habitat change for our species. The advances in biology and medicine compel us to rethink our approach to our own bodies, to reproduction, to sickness, to death [...] The development of nanotechnologies — that can produce materials of complex intelligence, microscopic artificial symbionts of our bodies and calculators of a capacity several magnitudes higher than today's ones — could turn upside down our attitude towards natural necessities and work. The improvements of the digitally supported cognitive prosthesis are deeply modifying our intellectual capacities, in the same way as mutations in our genetic patrimony would. The new techniques of communication through virtual worlds are restating the question of social cohesion. (Lévy, 2002: 16)

Pierre Lévy puts the stress on a type of nomadism that not only relates to physical places, but also to the abstract 'anthropological place' that, in terms of Marc Augé (1996), is a concrete yet symbolic construction of space. The anthropological space with which the nomad is confronted is mainly the space surrounding a single body (space of the individual), but is also the space where things and individuals interact (space of collective relations).

When Marc Augé refers to the new modern age and to its characteristic nomadism with the term *Supermodernity*, he is emphasizing its sense of excess: our time is overloaded with events, our space is compressed by faster and faster transport devices, our body sways in the midst of urbanization and new migratory flows. The body moves continuously from places to non-places. Non-places are the physical space-time products — from airports to big shopping malls or huge transport infrastructures — of a *glocal* nomadism that, in its more negative sense, has brought the individual back to a solitary condition.

The individuals that haunt the spaces of the non-places are nomads. Nomads because they assume in them temporary identities: identities restricted to the time of their visit and use of the non-place and mainly characterized by anonymity. Such provisionality of the identity does not escape to solitude and similarity, but can nevertheless bring some pleasure to the individual.

> Victim of a mild possession to which he abandons himself with more or less desire and conviction, he takes some pleasure — like any one possessed — in the passive joys of dis-identification and the more active delight of playing a role. (Augé, 1996: 94)

The non-place complies with the functions of the socio-technical system but acts simultaneously as a platform for new forms of freedom. The nomad of the non-places is in transit — where transit is the opposite of sojourn and residence — but not like a traveler — a travel is not important in itself but because of its destination —, rather like a passenger of a trip without a final goal.

The nomad interacts with the house (the home) and its inside and outside, with the law and its restrictions to freedom, with the place and its limits, with the non-place and its spaces of dis-identification, and even with the technologies, which constitute his new dwelling.

And nowadays, what other somewhere-else could the nomad stroll to? Apparently he cannot stray beyond the limits of law and rule because the principles that monitored these are in crisis — State, family, work. Today's somewhere-else seems to simply correspond to all possibilities, to a generic 'could be' (Maffesoli, 2000) that progresses in a permanent tension of wide-spread aspirations aimed at new types of identity search — from the new religious fundamentalisms to the techno-antagonistic subjectivities.

Squeezed between particular/specific/local and global, the individual is a nomad that goes astray towards a new aperture and a social system that is interpreted as collective. The union between individual nomadism and collective spirit, typical of the glocal movement, is a mirror image of the natural bond between the nomad and the tribe[1]: as a matter of fact, errancy implies the setting up of a survival strategy, necessarily based on conducts of mutual help and solidary support.

All in all, the body swings between two destinies: embodying the very last reassuring home, or being inevitably a techno-body that has freed himself from its original territories and wanders in search of new quarters. On one hand, the techno-body is a nomad in relation to the limits of

---

[1] Used here in the sense given by Howard Rheingold, (1993): virtual communities = real time tribes. The term comes back in Mario Ricciardi's critical essays (1998), where he deepens its meaning, linking it with the techno-social contemporary times.

biology; on the other hand, it is the vehicle (the support) of nomadism through the refunded space of the information technologies. The techno-body is at the same time resistance and proposition.

I have chosen to describe the work of the artist Marcel.lí Antúnez Roca, Catalan techno-performer, because he proposes a nomadism with a subversive character. He uses a techno-physical mutation that is not aimed at showing the bewilderment in front of the machine, not even the bewilderment in front of life, but that is aimed on the contrary at the exploitation of the possibilities offered by the new technologies to construct a human being who can come, through his errancy, to himself.

Technology: Personalization and Intimacy

Technology is an extension of man that connects him to the world. That extension used to have a mainly instrumental function — according to Gehlen (1983) the role of technology is to allow mankind to survive and artificially make up for its limitations and organic shortcomings — but tends now to occupy a new dimension and become the wide-spread and pervasive (unavoidable?) habitat of the individual. Since McLuhan (1964), we are indeed aware of the close relationship between communication technologies and the human psycho-physical structure — according to McLuhan the role of technology is to allow the individual to overcome, through a self-amputation mechanism, the functional stress induced by accelerating rhythms and increasing labor burden, both historically imposed by the money Medium and the written-word Medium.

> If we agree on the point that technique is the essence of man, the first factor of legibility that changes in the age of technique is the traditional one according to which man is a *subject* and technique an *instrument* at his disposal. That might have been true in the antique world, where technique happened inside the city walls, that is, inside a *site* enclosed in nature [...] But nowadays the city has expanded until the confines of the earth, so that nature is reduced to a *site* in it, to an enclosed portion contained inside the city walls. Therefore technique is not an *instrument* in man's hands to master nature anymore, but the *environment* of man[2]. (Galimberti, 1999: 36)

---

[2] According to Galimberti technique is 'whether the *universe of methods* (the technologies) that construct altogether the technical system, or the *rationality* that characterizes their use in terms of functionality and efficiency. Because of these features, technique is not the result of human 'spirit', but the 'remedy' against human biological imperfection' (Galimberti, 1999: 34).

Technology has changed our way of knowing, our body structure and our environment by drawing closer and con-fusing, be it in practical or in theoretical terms, the dimensions that I will now sum up in dialectical couples: the concrete knowledge (or savage thought, in terms of Claude Lévi-Strauss) and the abstract knowledge (or scientific thought); the symbolic-reconstructive modality, representing the mind's knowledge, and the perceptive-motor modality, representing the body's knowledge; finally, on the technological level, the technologies of the mind (like the Artificial Intelligence of the 80ies), the technologies of the body and the so-called 'technologies of body/mind', to use Domenico Parisi's (1994) term (like Artificial Life an Virtual Reality).

The environment of the techno-bodies — the place where they are defined/located by concrete action, experience and contact — brings about and loudly asserts the necessity of a knowledge that shall not be exclusively based on a way of thinking domesticated by language and writing (the alphanumeric mind): on one hand, technology requires the body and not the mind to be apprehended; on the other hand, man is using more and more corporal technologies to apprehend the world around him. My comment is based on the analysis of the practices with techno-bodies, but it quite matches the theoretical debate of authors like Claude Lévi-Strauss (1962), Walter Ong (1982) — when he formulates the 'secondary orality' — or Jack Goody (1977) and his *domestication* model. Having ascertained the importance of technology in the processes of evolution/mutation of human thought, we consider henceforth the body as the place to apprehend the new technological instruments.

The 'knowledge of the body' is to be understood as the recovery of a perceptive-motor approach (sensitive, corporal, self-aware) in order to apprehend the new technological world in which we dwell — a world made of realities that, according to Derrick de Kerckhove's (1995) definition, are essentially tactile (e.g. Virtual Reality). This knowledge recovers the importance of the biological/corporal paradigm for the design of new machines (e.g. Artificial Life and Genetic Engineering).

I shall put forward here the expression *tactile orality* to define the particular knowledge of the body that emerges from the interaction between the body and the technologies of information. Tactile orality comprises, on the level of theory, both the studies that see in the new communication media a come-back of the usual dynamics in oral or pre-logical societies (Ong, 1982; Goody, 1977), and the studies that consider the sense of touch as the new place of knowledge (De Kerckhove, 1995). This theoretical approach is proposed as a *new domestication* based on the recovery of the experience of the concrete world, as Sherry Turkle (1995) also puts forward when she defines the archetype of the *bricoleur*. Mario Ricciardi, following Turkle's tracks, describes with these words the difference between the *bricoleur*, new model of the technological man, and the *ingénieur*, technological man of the modern era:

> The world of the *bricoleur* does not originate exclusively from the choice of mixing and from the 'do-it-yourself'; it is a proper intention, a will, a real action in front of both the physical and symbolic worlds. Technique is also a type of practice, a type of action not to be solely considered in its absoluteness and isolation [...] The *ingénieur*, on the contrary, is the offspring and the representative of a society structured on writing, written alphabetization and industrial labor: rationalization and physical control over the territory and society are his strategic duties. Logical-sequential thought is his mental structure. (Ricciardi, 2005: XV)

A body that knows is above all a body that feels — related to perception —, moves — related to spacialization — and speaks — related to temporality. Knowledge is first of all action, and only then reflection.

This vision of the world and of knowledge — that is, the concept of a living subject-body that gets to know by means of a combination between senso-motricity and symbolic capacity — has its source in the phenomenology of Maurice Merleau-Ponty (1945) and goes on through authors like Mark Johnson (1987) and George Lakoff (1987), who are closer to the cognitive sciences. It is a vision supported even in the field of neurosciences, in which Francisco Varela articulates the concept of *embodiment*: 'Knowledge depends on being in a world that is inseparable from our bodies, our language, and our social history — in short, from our *embodiment*' (Varela, 1991: 149). The notion of *embodiment* and that of knowledge as *enaction*[3] are real turning-points of Varela's analysis (1991, 1994): they represent the passage of the incorporated knowledge approach from the philosophical disciplines to the cognitive sciences, closely related to practical technological research.

All these considerations have opened new debates in the domain of technological research on Artificial Intelligence, which has produced some new viewpoints, like for instance those known under the denomination Artificial Life. From the 50ies and 60ies up until now, the design of 'intelligent' machines was principally based on the computational paradigm, according to which knowledge operates through logical rules of

---

[3] 'Knowledge creates, establishes, brings into being; it is a question of construction, no less than a question of reconstruction of something already existing. It is the activity of the observer that institutes the world, the world depends on the observer, it is inseparable from his structure' (Varela, 1994: 150).

symbolic manipulation; in order to translate them to a digital language, it was essential to imitate the intelligence of a highly skilled expert. This paradigm has been questioned in the last twenty years by connectionism and by the modelization of neural networks, which has led to dynamic, non linear, adaptative systems. The logics of sequentiality give way to the logics of parallelism and the brain becomes the model to imitate in the construction of new intelligent machines — the organic brain and its physical functions are the real seat of knowledge.

The outcome of such a transition has produced the shift from pure mind technologies[4] — those that have produced the modern computers — to 'mind/body' technologies, to use Domenico Parisi's terms (1989, 1994). Parisi is the main italian representative of the new field of research known as AL (Artificial Life). The mind/body technologies have in connectionism and in the model of the brain their starting point; they take into consideration, for the design of intelligent machines, the organism and its whole functionality.

> Artificial Life simulations not only emulate the nervous system of the organism, but also:
> • the rest of the whole organism, of which the nervous system is only a portion;
> • the environment in which the organism lives and with which it interacts;
> • the population of other organisms of the same species, of which that organism is a member;
> • the genetic material received by the individual from the parents at the moment of birth, which determines the fundamental properties of the organism.
> (Parisi, 2001: 157)

The technologies of Artificial Life are 'mind/body' technologies in the sense that they aim to recover the perceptive-motor modality and the knowledge of the body.

According to Francesco Antinucci, 'sooner or later in the future, we could witness a real historical 'nemesis' where the body would take its revenge on the mind: it will be precisely the technologies that the mind has produced ignoring the body that will make it possible not only to repossess it but even to expand it extremely wherever it had been blind, deaf or inert' (Antinucci, 1994: 24). In rather different terms, we can talk of a tactile destiny for technologies (McLuhan, 1964). Following the studies of McLuhan, De Kerckhove (1995) asserts that, in our technological present,

---

[4] 'Pure mind is the mind without a body, without a brain, without a hormonal or immune system' (Parisi, 1994: 133).

everything seems to revert towards the touch, everything deals with touch, particularly Virtual Reality, which transforms touch in a cognitive extension of the mind. The skin alters its condition: once a protective membrane, it becomes now an instrument of communication and cognition.

The *tactile orality* that I am writing about corresponds to a communication characterized by a secondary orality, or return orality (Ong, 1982) [5], in which the senses of sight and hearing — imperative in the society of the screen (Debord, 1967; Manovich, 2001) — lose their preeminence and are overcome by physical contact and proximity. The Society of the Spectacle — to use Debord's expression — gives way to the Society of Experiences.

Technology has become a portable, personal and household matter thanks to the component miniaturization, the hyper-development of calculation and the expansion of the interconnecting networks. The bulky, isolated machines have been replaced with small, weightless, interconnected ones. De Kerckhove uses the following terminology to depict the development of technology: from the inside to the outside (the body acquires prostheses and extensions with and within the technologies), from the outside to the inside (technology is psychophysically interiorized, like in wearable technologies or in bio-technologies) and then, again, from the inside to the outside (personal life becomes collective by means of the Net, which in terms of Pierre Lévy epitomizes the creation of a collective Mind and a collective Body).

The use of the adjective 'personal' in reference to the nature of the new technologies is due, on one hand, to personalization — the machines are build on the user and can, therefore, be personalized by him — and, on the other hand, to the intimate relationship between technology and the single person: technologies concern the very personal sphere of an individual that is in reciprocal connection with the world. Consequently, we are not talking about an individual 'personal' but about a collective 'personal'. Even when I refer to 'intimate' technologies — meaning those that are so physically close to the user that they become a part of his intimacy —, I'm speaking of an intimacy shared, be it willing or not, with the rest of the connected world. I use the term *intimate technologies* to allude to those worn on or under the skin, right on the limit between the inside and the outside — though being in contact with the exterior, skin is still intimate —: *wearable computers* and *cyborgs*.

---

[5] 'Secondary' in the sense that it does not mean a return to pre-literary communication: we are faced with a communication that includes its own historic evolution and is the result of a *re-mediation* process (Bolter and Grusin, 1999).

Wearable computers[6] are 'on' the body and connect the individual with an environment that may be surrounding or distant, that is why they belong to the broader category of *wearable environments*:

> The vectors of this research include the following:
> • *information in places*: media linked to location.
> • *smart rooms*: environments that sense inhabitants and respond to them.
> • *digital cities*: adding information capabilities to urban places.
> • *sentient objects*: adding information and communication to physical objects.
> • *tangible bits*: manipulating the virtual world by manipulating physical objects.
> • *wearable computers* or *smart clothes*: sensing, computing, and communicating gear worn as clothing.
> (Rheingold, 2003: 84)

The *wearable computers* are peripherals fixed on the body and provided with input devices — hand joysticks or arm controls — as well as output ones — displays or glasses that allow the screening of virtual images right in front of the eyes. They have the three following properties (Mann, 1998): *constancy* — they are always ready —, *augmentation* — they do not substitute our body but bring it instead complementary possibilities — and *mediation* — they intercede between the body and the environment, they may indeed protect our privacy but keep us enclosed in a solipsistic world.

The *ubiquitous computing* technologies are those that can be spread all over the environment. In that case, their simplicity and, often, their invisibility are the features that facilitate their effective employment by a large audience (Norman, 1999)[7]. Alternatively, the computers can be placed on the individual's body, miniaturization is then their main feature.

---

[6] There are many examples: from the first *dataglove* of T. G. Zimmerman, through a head-mounted display for augmented reality, to common use objects like watches, glasses or cellular phones. For an in-depth research on *wearable computers* refer, among others, to the following studies: those carried out at Boston's MIT (http://www.media.mit.edu/wearables/ — consulted January 2007); those known under the name PAN and carried out by Zimmerman's team also at the MIT (http://www.media.mit.edu/physics/projects/pan/pan.html — consulted January 2007); and finally those carried out by Steve Mann at the University of Toronto, like the Wearcam and Cyberman projects (http://www.cbc.ca/cyberman/ — consulted January 2007).

[7] Norman is well known for his theories on user-oriented design as opposed to technology-oriented design. In his book 'Invisible computer', he puts forward the theory that the computer must be invisible, that is, disappear into the object and

On the way leading from the *wearable computers* to the *computer implantations*, the utmost figure in the evolution of intimate technologies is that of the *cyborg*[8].

Marcel.lí Antúnez Roca

Marcel.lí Antúnez Roca's Performativity

The concept of the *cyborg* takes us to a particular case study: the artistic techno-body performances. I see in them a useful research field on some anomalous but indeed forerunning uses of technology.

A performance is, in fact, a thought in action. It is idea and action simultaneously. It is a processuality open to improvisation and experimentation. It often investigates and handles both reality and the imaginary visions of it, altering or even illuminating the core of human nature — body, life, love and death, individual and social identity. It is interdisciplinarity and concrete multimediality. To sum it up:

• Very often a performance repeatedly destroys the codes and the common practices; it can thus be considered as an act against the specificity of language.

• A performance is an excess of energy that enhances the asymmetry between meaning and references; in terms of Jean-François Lyotard (1973) it is 'théâtre énergétique'.

• As a creative action, the performance almost never stands up to expectations: it is unpredictable in its end result and subversive in its intentions; this unfulfillment makes it basically mobile and exceptional, which Richard Schechner (1984: 177) conveys with these words: 'the performative activities are fundamentally *processual*: i.e. a part of them

---

hide its complexity to the user's eyes.

[8] The term *cyborg*, coined by Clynes and Kline in 1960, stands for '*cybernetic organism*' and indicates the fusion of human and synthetic parts. On the origins of this term and the context of its appearance, refer to the founding text on cybernetics by Wiener (1948); for a review on the word's origins, refer as well to Tomas (1999); for a political view on the *cyborgs*, refer to Haraway (1991).

will always remain subject to transformation, and will be absolutely impossible to define'.

The performances with techno-bodies have the same nature as the happenings: they are *subversive*; they hurt the artist's body and the public's emotions; and they seek destabilization through excessive ways of expression. This type of performances brings to light that maximum of excess that aims to generate absolute alterity — the 'otherness' of a bioinorganic being.

Orlan, Marcel.lí Antúnez and Stelarc are among the most active and well-known artists in the field of new technologies. In these pages I shall review the work of Marcel.lì Antúnez Roca (Moià, Barcelona, 1959) because, for me, he embodies like no other the concept of the modern *bricoleur* — combination of craftsman and cybernetic man —, son of the *new domestication* that the new technologies have induced.

The specific field of action of Marcel.lì Antúnez is a theatrical performativity typified by interdisciplinarity and multimediality. At the heart of this field, technologies are a platform of controversy and, simultaneously, instruments at the service of a new sort of show where the digital features — real time interactivity, adaptability, database logics — submit to a completely new narrativity — resplendent with juxtaposition, synchronization, interaction and real-time hypertextuality.

I would distinguish four major phases in this artist's long career:

• The tragic total work of art of the 80ies collective performances, going from the Fura dels Baus — theater group cofounded in 1979 — to Los Rinos.

• The uncanny (*das Unheimliche*) in the first individual and technological works, going from *JoAn, el hombre de carne* (1992) to *Epizoo – Mueve NO toques* (1994).

• The theatrical techno-mutant machine of the exoskeleton performances, going from *Afasia* (1998) to *Pol* (2002).

• The lectures / mechatronic lessons: *Transpèrmia* (2003) and *Protomembrana* (2006).

From theater to installations, from performances to theater, from lectures to workshops, Marcel.lí Antúnez generates a global and cyclical work of art where the cycle never means a cunning element repetition but an expansion of the original premises and key problems. All this implies a courageous quest. A quest aiming at the construction of a new cosmogony, at the heart of which the techno-body stands as a solid life project.

The main activity areas of this artist's work can be summarized with the following fields: computer programming; interfaces and robotics — Antúnez has designed an *exoskeletral* robotical/corporeal interface to give creative mobility to his own body on the stage —; multimedia databases — images, sounds, video recordings that make up the narrative units in the composition/development of his performances —; video and audio creations — producing visual and musical *landscapes* —; sound and lighting design.

In my opinion, there are three works in this long career that epitomize the birth, the development and the maturity of Marcel.lí Antúnez's exceptional, powerful, and vigorously creative techno-body. I shall now describe them.

JoAn, el hombre de carne

1992 is the date of the first creation to openly deal with the questions of man/machine/robot and of technology as a creative power put at the service of the artistic dimension of performance: *JoAn, el hombre de carne* — literally *JoAn, the meat man* — is a real scale dummy with male human features, made of pork meat and mechanically articulated. It is a robot that reacts to external sound stimuli with movements of its own body. Its appearance calls to mind the relationship between the body and the medieval executing techniques, the sacrifice body, the Inquisition, the torture practices and the helplessness in front of violence. In fact, *JoAn* is more than an puppet in the public's hands — which already bestows it with the fascinating magic of being non human but able to move like a living creature. It is an automaton deprived of any content and exposed to an environment that turns into a torture place. Like any XIXth century automaton, *JoAn* is an artifact for entertainment, born nevertheless from a critical reflection on robotics in the contemporary age.

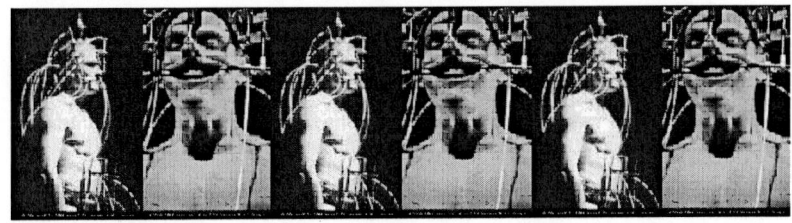

Epizoo, Mueve NO toques - Epizoo, Move DON'T touch

With *Epizoo, Mueve NO toques - Epizoo, Move DON'T touch*, Marcel.lí Antúnez launches in 1994 the first action to be named '*mechatronic interactive systematurgy*'. It is an action based on the full-blown use of mechanical, electronical and digital technologies on the theater stage. Antúnez stands half-naked on a plinth, surrounded by transparent tubes that connect pneumatic servomechanisms to a series of tiny orthopedic devices attached to various parts of his body. On one side of the stage stands a personal computer, on the screen of which a graphical interface shows an infographic simulation of the artist's body; by directly clicking on it, the public sets in action the orthopedic instruments that torture Antúnez's body as they maneuver his mouth, nose, eyes, ears, pectoral muscles and buttocks.

Epizoo is the exhibition of a glorious, excruciated techno-body, martyrized through an interaction that fills the gap between the virtuality of the images and the reality of the body. Antúnez exposes and discloses the nature of virtual reality: the place where body and world are to meet, the new pitch for man-to-man 'fights'.

Afasia

Afasia (aphasia), created in 1998, is the outcome of the artist's long search of a techno-body that could be a platform both for the new creative possibilities and the narrative discourse. *Afasia* is an interactive mechatronic performance based on the myth of the Odyssey. It unfolds as a genuine and absolute phantasmagory, which in its historical origins corresponded to a succession of images screened from a magic lantern. The Odyssey is an excuse to develop an interactive narration where Antúnez, equipped with an exoskeleton that is connected to the performance control system, triggers by his corporal movements the progression of the story — he activates in real time the various audio-visual landscapes that build up the scenic setting. The spectator can thus navigate through psychedelic settings that contradict the linear logic of the account and reorganize themselves in a hypertextual succession.

As in the pre-cinema magic lantern shows, Marcel.lí Antúnez's performances come down to a real-time selection and a composition (Manovich, 2001) of fantastic, visionary, magic elements where the techno-body is the motor and the *manager* of these new forms of Narration.

Marcel.lí Antúnez transforms the semiotics of the digital text into theater forms: structure multilinearity, content flexibility and recomposition, interactivity, spacialization and atemporality of the tale. It is the artist's techno-body, motor of the happening, that produces that transformation: the techno-body is forced to find a new *domestication*, to learn how to move in the digital parameters, to interiorize and naturalize the state of exception in which it is immersed by the use of the technologies, in search of a new theatrical scene.

> This 'horizontal' network enables us to disobey the hierarchical structure imposed by the market, so that we can reinvent the history of art and other forms of classification, thus becoming more permeable. (Marcel.lí Antúnez, 1998: 67)

The evolution of Marcel.lí Antúnez corresponds to a task of multidimensional and non-linear but methodical research: he aims to unveil (not describe) the body in our contemporary age. With *JoAn* he has dealt with the birth of a creature. With *Epizoo* that creature grows up, encounters/disencounters *the others* and engages in the struggle for life — dominating ones/dominated ones/executioners/victims. With *Afasia* the creature begins an initiatic trip, a navigation that intends no disembarkations but only the discovery of places.

The thought of difference (Perniola, 1997) and the Freudian concept of the *uncanny* experience linger in the background of all of his works. The thought of difference is the opposite of the thought of conciliation, based on the notions of harmony and neutrality. The thought of difference is opposed to a reconciled experience/existence, it is indeed anti-comforting, but it is not a thought of opposition; it rather responds to the logics of the 'AND' than to the logics of the 'OR'. The final result is the re-construction of a being that we could define, using Klossowski's terminology, as the simulacrum of the new techno-mutative era: 'Nevertheless, simulacrum is not to be understood as falsehood or lie, it has to be understood as something beyond true and false, as the gate to a third dimension where imaginary and real coincide or tend to coincide' (Perniola, 1997).

The aim embodied by Marcel.lí Antúnez is romantically a humanism and paradoxically an anti-cyborg, in the sense that he rejects the separation of mind, body and technology. He represents a state of exception because his procedures transgress biology as well as technology. We are not in a time for debating organic/inorganic, nature/culture, man/machine; this exception has become the rule. Our time is the time of resistance against that rule and the time of propositive disruption. Antúnez's body carries us from the cyber-body seen as a culture and media object to the techno-body understood as a critical, yet again powerful life project.

In our socio-technological context, Antúnez's art is, excuse me for the play on words, an exemplary example; it does not become symbol and it does not seal itself off in a frozen effigy — be it real or virtual — because it never betrays its own nature as a processual practice.

The power of that practice consists in Antúnez's ability to transform an art of resistance — resistance as the only possible approach to a reality that escapes us in every possible direction — into an art of regeneration. He transforms himself/ourselves by practices for life and vitality that do not deny contradiction, but do not accept them either in a post-modern way.

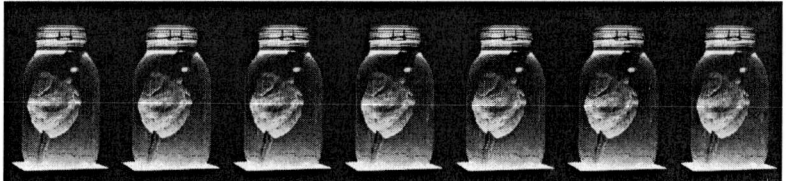

La vida sin amor no tiene sentido (Life without love has no sense)

Antúnez reminds us the tragical side of things by means of an exercise of revelation that uses the dionysic approach, the grotesque, the submission to horror and the fascination of the monster. To unveil but not to describe existence, that is what allows us to escape the rules and to recover our strength in exception.

> The Nymph Eco tried to win his love with fragments of his own speech, but in vain. He [Narcissus] was numb. He had adapted to his extension of himself and had become a closed system. (McLuhan, 1964: 51)

Antúnez's cyber-body recalls us the strength of a body already lost but alive under the ashes of our memory. It proposes us to come back as masters of the rule, by way of the exception of a post-digital and biotechno-logical body: *unrestrictive, unmonopolizing, controllable, communicative* (Mann, 1998).

> My body is continuous, not merely a support for my brain, my mental I, but a unique being in which mind and body form a single thing, pulsion of death. The body has a thousand and one different facets: a wounded, pulsating, obscene, phantasmagorical, mechanical body. (Marcel.lí Antúnez, 1999: 198)

Conclusion

> It is the story of a society that is crashing down... and, while crashing
> down, it says over and over again, to gather courage... everything ok until
> now... everything ok until now... everything ok until now... the problem
> is not the fall but the landing.
> (Off-screen voice in the film *La Haine*, directed by Mathieu Kassovitz,
> 1995)

The unstoppable movement of technologies, and therefore of the techno-
bio-logical system (mankind, world), is the dramatic rule that brings us
inevitably to reposition ourselves. And this because, as Kassovitz puts it,
the problem is not the movement — falling or soaring? — but the stopping
point: the idiom 'where are we getting at?' becomes the anguished
question of contemporary society, a society since long far away from the
radiance and illusions of both modernity and post-modernity.
The repositioning begins with the body, primary substance of any change.
As we have seen, the body has become the place of knowledge in our
techno-logical contemporary era. And this incorporated knowledge finds
in corporal technologies — that reflect the evolution of contemporary
technology and that I have here defined as personal, intimate and tactile —
their own narcissistic, fascinating and numbing mirror (McLuhan, 1964).
How shall we wake up from the torpor induced by the new technologies?
How could we accept/master the movement not as a decay but as a
horizontal succession of happenings/events that constantly shift the
boundary limits and accordingly the exception?
    Performativity is the dimension that we should try to master, because it
is the binding place, the place where technologies and corporeity meet.

Performativity is action, experience and movement, three words that describe very well the processual nature of the new communication technologies — not anymore instruments designed according to the standards of modernity, but places governed by flexibility and permanent reorganization — as well as the processual nature of the constitutional mobility of the body — the body does not stop, it is born, it grows, it dies, the body moves, it is never representation but eluding instant and unstoppable event.

That is why I have proposed the Arts as the research field. They are the place of performativity by definition, but also the place of the metaphors, where we can find again the strength of mind that will take us, far beyond the description of the state of things, to the realm of revelations — starting point for the construction of a new, concrete and real project of life. Techno-performative Art becomes thus the place of premonitions, of uncovering, of unexpected technology uses: in short, the place of the exception that can become a new rule. 'The artist picks up the message of cultural and technological challenge, decades before its transforming impact occurs' (McLuhan, 1964: 70).

And then again, how can we make the passage from Art to life, and from the exception of the 'stage' to everyday reality? I have put forward Marcel.lí Antúnez Roca's work because I think that his 'way' can possibly help to go from the NEW scenic reality to the NEW social one. Marcel.lí Antúnez is a committed artist, I would even say a down-to-earth one: his feet are on the ground and his body is in the substance of things — be it robotic, electronic, digital or bio-technological. The NEWNESS that we owe to him comes from his systematic, anticipating and original trajectory: from the newness of the Fura dels Baus, to the newness of the technological performances, and to the newness of the hypertextual narration. His techno-body goes away from the metaphors of Art and achieves through actions/revelations the reappropriation of the technologies. His goal could be a goal of ours: to master the digital rule in order to transform the techno-body from an anguishing evidence to a place of reappropriation of one's own history, using the strength of an incarnated tale where necessity and pleasure are in balance.

The artist can show us how to 'ride with the punch', instead of 'taking it on the chin'. (McLuhan, 1964: 71)

References

Antinucci, F. (1994) 'Il corpo della mente', in P.L. Cappucci, *Il corpo tecnologico*. Bologna: Baskerville.

Arendt, H. (1995) *Che cos'è la politica*. Milan: Edizioni di Comunità.

Augé, M. (1996) *Nonluoghi, introduzione a una antropologia della surmodernità*. Milan: Elèuthera. (Orig. pub. *Non-lieux, Introduction à une Anthropologie de la Surmodernité*. Paris: Seuil, 1992.)

Balsamo, A. (1995) 'Forms of Technological Embodiment: Reading the Body in Contemporary Culture', in M. Featherstone and R. Burrows (eds), *Cyberspace, cyberbodies, cyberpunk: cultures of technological embodiment*. London: SAGE Publications.

Benn, G. (1992) 'La conquista', in *Cervelli*. Milan: Adelphi.

Biocca, F. (1997) 'The Cyborg's Dilemma: Progressive Embodiment in Virtual Environments', in *Journal of Mass Communication (JCMC)* 3(2), URL (consulted January 2007): http://jcmc.indiana.edu/vol3/issue2/biocca2.html

Bolter, J.D. and Grusin, R. (1999) *Remediation: Understanding New Media*. Cambridge, MA: The MIT Press.

Bricken, M. (1991), in M. Benedikt *Cyberspace: First Steps*. Cambridge: The MIT Press.

Castells, M. (1996) *The Rise of the Network Society*. Oxford: Blackwell.

De Kerckhove, D. (1995) *The Skin of Culture: Investigating the New Electronic Reality*. Somerville Press.

—. (1994) 'Remapping sensoriale', pp. 45-61 in P.L. Capucci (ed.) *Il corpo tecnologico*. Bologna: Baskerville.

Debord, G. (1967) *La societè du spectacle*. Paris: Buchet-Chastel.

Galimberti, U. (1983) *Il corpo*. Milan: Feltrinelli.

—. (1999) *Psiche e techne, l'uomo nell'età della tecnica*. Milan: Feltrinelli.

Gehlen, A. (1983) *L'uomo. La sua natura e il suo posto nel mondo*. Milan: Feltrinelli. (Orig. pub. *Der Mensch. Seine Natur und seine Stellung in der Welt*. Berlin: Junker und Dünnhaupt, 1978.)

Goody, J. (1977) *The Domestication of the Savage Mind*. Cambridge University Press.

Haraway, D. (1991) *Cyborgs, and Women: The Reinvention of Nature*. New York: Routledge.

Hayles, N.K. (1996) 'Virtual Bodies and Flickering Signifiers', in T. Druckrey (ed.) *Electronic Culture, technology and visual representation*. New York: Aperture.

Johnson, M. (1987) *The Body in the Mind. The Bodily Basis of Meaning, Imagination, and Reason.* Chicago: University of Chicago Press.

Kammermeier, P. and Kron, A. and Hoogen, J. and Schmidt, G. (2004) 'Display of Holistic Haptic Sensations by Combined Tactile and Kinesthetic Feedback', *Presence: Teleoperators and Virtual Environments* 13(1).

Lakoff, G. (1987) *Women, Fire, and Dangerous Things: What Categories Reveal About the Mind.* Chicago: University of Chicago Press.

Lakoff, G. and Johnson, M. (1999) *Philosophy In The Flesh: the Embodied Mind and its Challenge to Western Thought.* New York: Basic Books.

Lévi-Strauss, C. (1962) *La pensée sauvage.* Paris: Librairie Plon.

Lévy, P. (2002) *L'intelligenza collettiva.* Milan: Feltrinelli. (Orig. pub. *L'intelligence collective - pour une anthropologie du cyberspace.* Paris: La Découverte, 1994.)

—. (1995) *Qu'est-ce que le virtuel.* Paris: La Découverte.

Lyotard, J-F. (1973) 'La Dent, la Paume', in *Des dispositifs pulsionnels.* Paris: Union Général d'Éditions.

Lombard, M. and Ditton, T. (1997) 'At the heart of it all: The concept of presence', *Journal of Mass Communication (JCMC)* 3(2), URL (consulted January 2007): *http://jcmc.indiana.edu/vol3/issue2/lombard.html*

Maffesoli, M. (2000) *Del nomadismo. Per una sociologia dell'erranza.* Milan: Franco Angeli. (Orig. pub. *Du nomadisme. Vagabondages initiatiques.* Paris: Le Livre de Poche, 1997.)

Mann, S. (1998), 'Wearable computing as means for personal empowerment' (Keynote Address for The First International Conference on Wearable Computing, ICWC-98), URL (consulted January 2007): *http://wearcam.org/icwc/empowerment.html*

Manovich, L. (2001) *The Language of New Media.* Cambridge, MA: MIT Press.

Marcel.lì Antúnez Roca, (1998) catalogue *Marcel.lì Antúnez Roca, performances, objetos y dibujos.* Barcelona: Mecad.

—. (1999) catalogue *Marcel.lì Antùnez Roca – Epifanìa.* Madrid: Claudia Giannetti ed.

McLuhan, M. (1964) *Understanding Media.* New York: Mentor.

Merleau-Ponty, M. (1945) *Phénoménologie de la perception.* Paris: Gallimard.

Nancy, J-L. (1995) *Corpus.* Naples: Cronopio. (Orig. pub. *Corpus.* Paris: Métailié, 1992.)

Norman, D.A. (1999) *Invisible Computer: Why Good Products Can Fail, the Personal Computer Is So Complex and Information Appliances Are the Solution.* London: MIT Press.

Ong, W. (1982) *Orality and Literacy: The Technologizing of the Word.* London and New York: Methuen.

Parisi, D. (1989) *Intervista sulle reti neurali: cervello e macchine intelligenti.* Bologna: Il Mulino.

—. (1994) 'Tecnologie della mente/corpo', in P.L. Cappucci (ed), *Il corpo tecnologico.* Bologna: Baskerville.

—. (2001) *Simulazioni, la realtà rifatta nel computer.* Bologna: Il Mulino.

Perniola, M. (1997) 'Il pensiero della differenza', URL (consulted january 2008): http://www.hackerart.org/media/amc/perniola.htm

Rheingold, H. (1993) *The virtual community: homesteading on the electronic frontier.* Reading, Mass.: Addison-Wesley.

—. (2003) *Smart Mobs: The Next Social Revolution.* Cambridge: Perseus Publishing (First edition Place: Basic Books, 2002.)

Ricciardi, M. (1998) 'Le comunità virtuali e la fine della società testuale', in P. Ceri and P. Borgna (eds), *Tecnologie per il XXI secolo.* Turin: Einaudi.

—. (2005) 'Lo schermo e lo specchio', in S. Turkle, *La vita sullo schermo.* Milan: Apogeo.

Schechner, R. (1984) *La teoria della performance 1970-1983.* Rome: Bulzoni editore.

Stone, A.R. (1996) *The War of Desire and Technology at the Close of the Mechanical Age.* Cambridge, MA: MIT press.

Tomas, D. (1995) 'Feedback and Cybernetics: Reimaging the Body in the Age of Cybernetics', in M. Featherstone and R. Burrows (eds), *Cyberspace, cyberbodies, cyberpunk: cultures of technological embodiment.* London: SAGE Publications.

Turkle, S. (1995) *Life on the Screen: Identity in the Age of the Internet.* New York: Simon and Schuster.

Varela, F. and Thompson, E. and Rosch, E. (1991) *The Embodied Mind: Cognitive Science and Human Experience.* Cambridge MA: MIT Press.

—. (1994) 'Il reincanto del concreto', in P.L. Capucci, *Il corpo tecnologico.* Bologna: Baskerville.

Wiener, N. (1948) *Cybernetics: or control and communication in the animal and in the machine,* New York: John Wiley.

Weiser, M. (1991) 'The Computer for the 21st Century', in *Scientific American,* 265(3): 94-104.

TOURISM IN CONTEXT: THE TOURIST'S CYBERBODY

GIULI LIEBMAN PARRINELLO

Abstract: Featured by mobility and complexity and evolving at a very rapid pace, tourism, strictly intertwined with globalization and technologies, is a privileged research field, as its main social actor, the tourist, can be seen as the representative of the contemporary shrewder individual. In the last few decades, featured also in the tourism phenomenon by a post-Cartesian attitude, we have witnessed a transition from sightseeing to a more complex multisensory body experience.

The article considers at first some mainstream theoretical suggestions from the field of tourism social studies (authenticity, gaze, experience) and moves from a cultural anthropological evolutionary approach to human beings and technologies, and from the recent perspective of neurosciences, which is compatible with choice processes bound with language, memory and the categories of mind.

An immense tourist 'imaginary' has developed, ranging from myth to rite to true cybertechnologies with inter-active participation and the overlapping of the real and the virtual intertwining. From a more psychological than anthropological view, the pursuit of happiness typical of the tourist is envisaged. Encouraging suggestions from a recent research field of cognitive psychology coupling economics and happiness are taken into consideration and the concept of happiness is seen as strictly interwoven with complex memory and language processes.

Keywords: Tourist, body, technologies, neuroscience, 'imaginary', happiness, subject

Introduction

The focus is on the tourist and the wholeness of her/his mind and body. The tourist is no longer seen as the cultural nitwit of mass tourism, but the representative of the contemporary shrewder individual confronted with new horizons and new challenges (Graburn 2001, Liebman 2001), not only in the highest and most extreme forms of her/his experienced circular mobility, but also in the quiet aspects more akin to leisure.

Firstly the mainstream theoretical suggestions from the field of tourism social studies are reviewed (MacCannell, 1976; Urry, 1990; Cohen, 2004), mainly uninterested in the body, but integrated with hints of a recent trend moving away from the 'gaze' and of some contributions with the accent on the body in the last decade (Veijola and Jokinen, 1994; Wang, 2003).

Secondly broad space is given to an interpretation of technologies considering them as 'extensions of man' (McLuhan, 1967) and coupling them with human culture (Gehlen, 1984; 1993). In fact technologies have necessarily gone hand in hand **with** the tourist's development, from the transport technologies of the industrial revolutions up to the inception of space tourism, to the most advanced ICTs interfering with her/his mind and body.

Then, thirdly, the neurobiological basis is outlined. Especially in its evolutionary meaning (Edelman, 1993; 2004) higher-order consciousness does not imply reductionism, it is compatible with choice processes bound with language, memory and the categories of mind. It is also consistent with suggestions coming from cognitive science (Lakoff and Johnson, 1980), justifying a 'universe of consciousness' (Edelman and Tononi, 2000).

Fourthly an approach to the embodied tourist subject and her/his language is tried through her/his complex 'imaginary', related to myth, wellness and happiness. Thanks to the concept of happiness (Enzensberger, 1996 [1958]), relevant suggestions from a recent research field of economics and cognitive psychology (Kahneman and Krueger, 2006) are taken into account. The accent is put on the remembering rather than experiencing subject, connecting these hints with memory and language processes. Thus we are also taken closer to a specific perspective of narrative (Bruner 2005), where the sovereign tourist subject consolidates her/his embodied experience and structures it, embracing the whole range of human possibilities.

The Conceptualization of the Tourist in Tourism Social Studies

The scientific basis of tourism studies can probably be ascribed to the founding fathers Hunziker and Krapf (1942), who, as forerunners, gave an organic definition of tourism.

The still usual formulation as 'a sum of phenomena and relationships' (McIntosh and Goeldner,1995:10) indicates the magnitude and complexity of this economic, social, environmental etc. aspect of contemporary society. After a productive blossoming in the 80s and 90s (Lanfant, 1995), the proper focus on a sociological concept like tourism has apparently already been abandoned by some scholars in favor of related, possibly more encompassing, contemporary phenomena like mobility (Rojek and

Urry, 1997; Sheller and Urry, 2004). Certainly it cannot be ignored that, as Cohen puts it, .there is 'the threat of the 'end of tourism' as a distinct type of activity' (Cohen, 2004: 317); moreover, the menacing 'loss of boundaries between tourism and leisure' (Cohen, 2004:8) emerges from the core of tourism itself.

And yet tourism reflects the challenges of our time and it is argued that not only should the tourism phenomenon be considered in the light of its specific historical genesis bound to the first industrial revolution, then through its developments to post-industrial society, but that just its evolution is a seismograph of also new and unforeseen features in the present globalization era (Meetham, 2001; Theobald, 2005). Even if de-differentiation is a fact, it is still consequential to deal with it from the point of view of tourism studies. Moreover, putting the tourist at the center of the phenomenon can still reserve some surprising results.

i) Mainstream theories

During the last three decades, coupling sociological with anthropological studies gave a fundamental contribution, but faced with the phenomenon's complexity, the pendulum swung several times between tourism and the tourist. It is worthwhile dwelling on some concepts, representing a partial link between tourism and the tourist, but especially regarding the tourist, which were particularly successful and sometimes indiscriminately adopted.

To simplify, we can outline on one hand the theory of 'authenticity' by MacCannell (1976), on the other the 'gaze' by Urry (1990), which for decades have monopolized the field of tourism social studies, in the sense that their theories '*became* general theories of tourism' (Franklin, 2003: 265). In between there is the more composite approach to the tourist experience, which is compatible with the former.

MacCannell's tourist (1976) has quite a different profile compared to the simpleminded human being outlined by Boorstin (1964). He is a modern creature in search of authenticity and is looking for it through contemporary sightseeing. Like a secular pilgrim, even if it is not expressly stated, MacCannell's tourist is in quest of the Other and of a way out of the limits of everyday life. The author is openly inspired by Goffman for the concept of front- and backstage and for the 'staged authenticity' produced by the inhabitants for the tourists (MacCannell, 1973).

The category of authenticity, which is object-related, has often been analyzed and criticized (Cohen, 1988; Olsen, 2002). Going beyond

MacCannell's sightseeing, typical of Western middle-class tourists, the Berkeley anthropologist Graburn, aware also of the crosscultural aspects of travel understood as pilgrimage (Graburn,1983; Graburn, 1995) has recently tried to 'relocate' the tourist (Graburn, 2001: 150).

Urry's 'gaze' is inspired by Foucault, but strictly bound to the history of British tourism development (Urry, 1990). The 'gaze' is explicitly not 'the single tourist gaze as such. It varies by society, by social group and by historical period. Such gazes are constructed through difference' (Urry, 1990:1). Urry distinguishes between a romantic gaze, typical of the educated middle-class, and the collective gaze, less properly centered on the eye, and somehow more British working-class.

The Tourist Gaze was written at a time when the sign and the symbolization of an object were becoming more relevant than the thing itself (Franklin 2003). Compared to MacCannell's somewhat romantic dream of authenticity, the world has in the meantime developed to more hyperreal aspects in the work of Urry, in a new prospect of modernity, later on dealt with in *Economies of Signs and Space* (Lash and Urry, 1994).

If we now go over to the third concept, related with the issue of tourist experience, relevant considerations come from a renowned sociologist like Bauman. From a more general post-modern point of view, Bauman takes a semantic attitude, and considers tourists and vagabonds as consumers who have a mostly aesthetic relationship with the world (Bauman, 1998: 94). Clearly, theorization in tourism studies goes beyond this anyhow suggestive and brilliant statement. Uriely's recent state of the art illustrates the wealth of this broad concept, stressing among the other points the passage 'from focusing on the toured objects to the attention given to the role of subjectivity in the constitution of experiences' (Uriely, 2005: 209) and the negotiation of the tourist's meanings.

It is worth remembering the seminal typology of tourist experiences by Eric Cohen, which is inspired by Schutz for the phenomenological aspect and by Eliade for the 'center out there' (Cohen 1979). In recent approaches, still based on his phenomenological articulation, Cohen now questions the search for authenticity, interpreting it as a social construct and underlines 'the playful enjoyment of 'surfaces'' having superseded the quest for authenticity as 'the cultural model of the touristic experience of 'postmodern' tourists' (Cohen, 2000a:216). In the strict sense, there is no apparent emphasis on physical involvement, which can altogether be considered implicit. Cohen distances himself also from the 'gaze' theory asserting that 'this argument applies primarily to sightseeing rather than vacationing tourism, where total sensual emersion rather than a dominant visual experience appears to be the rule' (Cohen, 2000b: 546).

Rich food for thought is given by the collective volume *The Tourist Experience* (Ryan, 2002). It offers a broad spectrum of tourist experience, open to all kinds of pleasure travels and to psychological aspects like motivation, memory, to the beach location, to fundamental structures like time. Although arguing with Pearce's allegedly inadequate concept of tourist experience (Pearce 1988), a definite and solid concept of experience is unfortunately missing.

Even though they don't focus primarily on the concept of experience, interesting suggestions come from the considerations of Wang *Rethinking Authenticity in Tourism Experience* (Wang, 1999). Here existential authenticity is proposed and classified into an intra-personal authenticity featured by bodily feelings and an intra-personal experience involving 'self making' or 'self identity'. Wang has dealt with authenticity and experience also in his *Tourism and Modernity. A Sociological Analysis* (Wang, 2000).

ii) The body as an epistemological challenge

As we have seen, in the last decades only hints of a more complete psychophysical discourse have come out especially if compared with tourist performances, be they traditional hedonistic tourism or new particularly pronounced forms or formulas like wellness, cocooning etc. on the one hand and adventure involving physical effort, 'adrenaline' tourism, on the other.

Tourism marketing has always played on the sensuous aspects of tourism. On the contrary, tourism theories have proved to be inadequate concerning the latter.

In 1994, a provocative essay (Vejiola and Jokinen, 1994) puts the problem of the body at the center of tourism research. Not only the absence of the body from sociological studies on tourism is highlighted, but also of the *writing subject*, the analyst supposedly lacking a body.

The feminist input is evident and the Finnish authors are, not by chance, acknowledged by Swain in the special issue of the 'Annals of Tourism Research' dedicated to gender in tourism (Swain, 1995). Swain significantly states that the body is emblematic of what is missing in universalizing social science theories in general, and in tourism studies in particular (Swain, 1995: 256).

This approach has been fruitful, and wider and more explicit claims have also been coupled with the inclusion of 'sexed and sexualized bodies in tourism' (Johnston 2001).

Accompanying the abovementioned approach to the body, the impression is given of a more general discourse following also other paths, as in the case of senses other than the visual gaze. Urry himself admitted that there was not only the gaze that counts. In a relevant interview, bearing the significant title *The Tourist Gaze and Beyond* (Urry, 2001) he conceded:

> I am thinking here of several developments that emphasize a closer, more sensual relation to tourist objects [...] (Urry, 2001:121).

It is intuitable that many objections to Urry have been raised (Morkham and Staiff, 2002; Quan and Wang, 2004). E.g. moving from New Zealand tourism reality, Perkins and Thorns (2001) reject the global application of the gaze, stating the importance of performance in tourism recreation. Dann and Jacobsen (2002) not only stress the importance of a sense like smell (*Leading the Tourist by the Nose*), but point out the tourist's polysensual completeness.

The beach is a privileged place, and in fact not by chance attention has been dedicated to the beach by several studies, having superseded the contempt for the components of the 'unholy quartet' of the four S's (Selwyn, 2000). Urbain (1994) and Hennig (1997) show interest in the beach as a scene for holiday life, and envisage the disappearance of traditional hierarchies in the bodies, considering this situation close to the liminal one of carnival.

Another approach is to connect body to nature in a very encompassing way (Franklin, 2003). The ecological movement, with nature and culture bound together again and the body placed in this context, is envisaged in the light of sports tourism, like surfing, climbing and also ecotourism in general. From different points of view Liebman (2001) and Wang (2003) take into consideration the vast, not only sociological, literature of the 80s and 90s on the body. Highlighting the technological body, Liebman stresses a common post-Cartesian attitude and the different aspects of body evaluation, where the tourist's body is seen as particularly privileged (Liebman, 2001). Wang (2003), in a mainly philosophical approach specifically dedicated to the body, follows Bourdieu in defining it as a body-field, and distinguishing four types of body-fields (Wang 2003, 129). The tourist body-field is to be seen as a 'permissive' one and the tourist's body as a potentially natural body compared to the extremely socialized body of the civilization process.

The Extensions of the Tourist

Very clever (though somehow implicit) suggestions emerge from Enzensberger's (1996 [1958]) comparison of two travellers, one on a traditional walking tour, the other travelling by air. The first described by the novelist Jean Paul is a poor-man's lawyer, travelling from Kuhschnappel to Bayreuth at the beginning of the nineteenth century and the other, by Max Frisch, a UNESCO engineer called Faber travelling from New York to Caracas around 1957. Enzensberger's intuition places the great phenomenon of tourism between the two journeys. We can try to complete this picture by coupling tourism with technologies: three Industrial Revolutions have come and gone between the rare trip on foot and the routine of air travel. Actually, to update the picture, one more traveller would be needed, in the person of a possible female space traveller. Speaking more in depth of tourism and of technologies, we should try firstly to get to the heart itself of technologies, although it is not an easy task, as anticipated by the title of this part.

The subtitle of McLuhan's *Understanding Media* is significantly *The Extensions of Man* (McLuhan, 1967). The renowned statement 'The medium is the message' has in fact more implications than are usually considered:

> This is merely to say that the personal and social consequences of any medium – that is, of any extension of ourselves – result from the new scale that is introduced into our affairs by each extension of ourselves, or by any new technology (McLuhan, 1967:15).

If these words are read carefully, it will soon be clear that McLuhan is linking up anthropological development, the media and technology in a global discourse. Actually, the anthropological connection is probably a bit more complex. Particularly relevant is the philosophical anthropology of Arnold Gehlen (1984; 1993) considering technologies as artefacts and as a mirror of human beings, like the human beings themselves *nature artificielle* (Gehlen, 1984: 13).

Gehlen, who takes the opposite view to Heidegger, is a fundamental reference point, although his influence is only implicitly acknowledged by many scholars. By way of simplification, following in the footsteps of Herder and Uexküll, he asserts that the need for technology is due to the imperfections of human beings (Gehlen, 1984; Gehlen, 1993). Technology can thus be considered in an anthropocentric way as a kind of orthopaedic prosthesis, which not only substitutes, but also increases the potential of human abilities. The human being is a tool-making man, and the world of technology can hence be considered a kind of 'Great Man'. Gehlen speaks of substitution, strengthening and relief of the human organs. They can also be systematized as locomotory, sensory and intellective prostheses (Maldonado, 1989; Maldonado, 1997; Liebman, 2001).

The first particular light on tourism is cast by locomotory prostheses, which regard all vehicles, from the bicycle to the airplane.

Some studies on the past offer valuable points for further research. Schivelbusch (1977) made an interesting analysis of the consequences of speed on the railroad in the 19th century and developed it into a study of the psychosociological changes resulting from technological innovation in the Industrial Revolution.

On the basis of historical achievements, McLuhan establishes the links between wheel, bicycle and airplane (McLuhan 1964) and then gives particular attention to the motorcar, which he calls by an expression which became famous later on 'The Mechanical Bride' (McLuhan 1964, 232).

Further discussing transport technologies it is Gehlen himself, considering the airplane, who states that it realizes the three above-mentioned principles in one, in the sense that it substitutes wings, it beats all the organic flying capacities and saves effort for those who want to visit distant lands (Gehlen, 1984). Technologies don't go ahead alone, they are intertwined. So air transport would not have been so successful for long-haul destination if it was not for the development, taking place in the meantime, of a computerized reservation system, which has also allowed very efficient low-price charter flights. Space tourism is the latest horizon of technological tourism, which up to now has come true for just a few persons but which is bound to develop intensively, as the potential market is estimated at 10 million people (Smith, 2001).

At this early stage, space tourism can be considered similar to the most extreme forms of adventure tourism, but it actually has some features akin to those of mass tourism (Cohen, 2004). In fact it is the technology, being all in all a transport vehicle, which makes the difference. The space-ship cabin is for collective transportation, where the space traveller, like the mass tourist, is wrapped in a kind of 'environmental bubble', precluded from leaving and confined mostly to the 'gaze', although to the very special 'gaze' directed at the earth from outside.

Speed has increased a hundredfold from the early railway trips to contemporary supersonic flights. As far as the coordinates time and space are concerned, Bloch wrote very stimulating notes about the compression of space into time. If the landscape changes very quickly, then the traveller perceives a transformation of space into time and of time into space. Time is filled like space and space becomes a medium of change, as only time can normally be. As a result there is a reversal of the usual order of perception, we have filled time in an apparently changed space (Bloch, 1959). The consequences for world social time became evident in the course of the 19th century. Between 1870 and 1920 – corresponding to the Golden Age of international tourism in the framework of an early process of globalization – the new world coordinates underwent a process of standardization but also of heterogeneization, as a result both of technological innovations and of multilateral negotiations (Kern 1988). Nowadays the concept of real time, practically corresponding to the elimination of distance, has become a common issue.

It is not only mobility itself, but its multiplication and serialization that counts. Thomas Cook made it possible, through a realization that could be rightly called 'Cookism' (Urry, 1994). The serialization and the standardization of mass travel are of paramount importance and full of implications, as already Enzensberger (1996 [1958]), stated, also without adding the dimension of hyperreality, later on theorized by Eco (1987) and Baudrillard (1983).

Among the second type of technologies, that is intensifying the activity of the organ, and being a sensory prosthesis, photography has played a fundamental role, as it has gone hand in hand with tourism since the 19th century.

> It is common knowledge that photography – developing at the same pace as tourism - has contributed to the increasing importance of the image. Photography can reach people everywhere and also individually. Through photos, transformed into posters, the most important destinations entered the tourism imagery. Photographs and postcards are still the mainstays of tourism practice and representation (Albers and James 1983, 1988, Cohen et al. 1993, Dann 1996). The tourist experience has also possibly been increased and enhanced by instant development [...] (Liebman, 2001: 210-211).

In this context the use of tourism brochures should be mentioned, still employed in spite of the more and more widespread use of the internet. The luxurious halo, the tangible suggestion of glossy paper, the 'bold and the beautiful' usually depicted give no sign of disappearing.

At this point, the internet can then be seen as the top of an intellectual prosthesis. From this point of view its function should probably be considered a kind of strengthening and intensification of the mental organon, but it is at the same time strictly intertwined with all the different embodied technologies.

Although 'cyberanthropology' is a very precise term, for many people it still represents a kind of threat evoking the futuristic and dehumanizing science-fiction perspectives of cyberspace. Here, by contrast, the aim is to trace the term back to its origins, thereby going on with the discourse stressing the continuity and evolution of human nature, culture and technology. Among the most recent positions of scholars interested in the anthropology of cyberspace and cybertechnology, Pierre Lévy emphasises that the progress of cognitive prostheses with digital support modifies human intellectual capacities in depth, as if they were mutations in their genetic patrimony. Moreover, he stresses the evolutionary process of human kind which is not only not yet accomplished, but where a sudden acceleration can be noted (Lévy, 1996; Lévy 2000).

Tourism is information-intensive and information-based (Poon, 1993; Sheldon, 1998), but the interest in tourism studies was mainly in the field of the industry in its different technical aspects, than on individual tourism, if not concerning the so-called distribution sector.

As Dann has ascertained, it is nowadays not only the positive response of potential tourists to the promotional appeals of the industry. It is the inception of a radically new communication mode substituting the traditional top-down, leading to a less monological and more dialogical or trialogical channel (Dann, 2005). The tourist's 'voice' is especially evident in the phenomenon of travel-blogging, which leads back to the cyberanthropological dimension and to collective intelligence (Dann and Liebman, 2005). Also new paths of identity emerge, to be considered as 'selving', in continual formation and reformation (de Kerckhove, 2005).

The controversy concerning Virtual Reality seems to have died down, after years of intense discussion polarizing it (Williams and Hobson, 1995; Prideaux 2002). At first confused with the internet, proper VR is in fact not so easy to practise. Today true cybertourism is seen as an opportunity for marketing, for museum exibitions and/or in order to protect sensitive sites.

As Prideaux puts it,

In the coming post-contemporary era, the tourist gaze will enable participants to be removed from the physical constraints of structures that currently define the boundary between the authentic and the contrived to move into a cyberworld where voyeurs will become participants able to write their own scripts for the touristic experience they enjoy. However, the process of converting what can be described as today's science fiction into science fact is at least several decades away and will depend on new technologies (Prideaux 2002, 333).

Probably nano-technologies will decide the future of these developments, where the cyber-scientific prospect might be the worrying blurred distinctions between the human brain and technologies.

Furthermore, more obvious technologies should be mentioned, from garments, sports equipment to food preservation, helping not only the most extreme forms of adventure tourism, but also quiet family camping-site activities.

Also technologies as attractions are very frequent, from naif 18th century factory preservation to major technical realizations, like towers, dams, urban systematizations to the cathedrals of tourism itself, that is innovative hotels.

Last but not least theme parks and entertainment complexes should be mentioned, in their emblematic dazzling but also more subtle composite function.

Technologies are not the main attractions of these establishments, from Disneyland to West Edmonton Mall to Las Vegas, but are one of their components, and their realization would have been anyway impossibile without the most up-to-date levels of hi-tech. For some aspects they bring to mind the world exhibitions of the 19th century with their technological content (Hennig, 1999), for others there is a strong appeal to consumerism. This confirms the de-differentiation between tourism and leisure (Cohen, 2004), also because theme parks have progressively expanded on the territory. Indeed, even in the apparent banality of these 'ephemeral spaces' (Minca, 1998), all the complexity of the tourist stock-in-trade is concentrated there, with its multiple different 'imaginary worlds' (Appadurai 2001).

Possible Answers of Neuroscience

Even if the contemporary mind-and-body debate and neuroscience can seem at first a far cry from tourism research, these issues can contribute to clarifying the tourist's mind-body relationship and her/his embodied experience.

In fact until recently the mind-body problem was a philosophical topic, outside the realm of empirical science. Nevertheless, today, this issue has finally been fully recognized as the encounter of different traditional disciplines like philosophy, psychology, physiology, biology, and of new scientific disciplinary areas like neuroscience, cognitive science and AI, which interplay with traditional philosophical and psychological sub-sectors. But it is especially neuroscience which represents the supporting substratum and that should be considered as a point of no return.

The current debate involves very authoritative scientists, names like Damasio (1994; 2003), Edelman (1993; 2004); Lakoff and Johnson (1980), Varela et al.(1996), Searle (1983; 1997); Clark (1997); Dennett (1991), etc.

Broadly speaking, Cartesianism has been superseded. Damasio's *Descartes' error* (1994) is often quoted because he proposes a joint vision of the individual, as a union of reason, emotions and feelings. The latter are then referred to a philosophical forerunner like Spinoza (Damasio, 2003) somehow prefiguring modern neuroscience.

Especially in its evolutionary Darwinist version advanced by Edelman, neuroscience is the latest frontier to confront. Neurons are nerve cells of the central or the peripheral nervous system and are mostly concentrated in the cortex, that is, in the outer part of the brain particularly developed in superior mammals, where higher-order functions are located. Following Edelman's studies, about 30 billion neurons are present in the cortex, whilst the number of their possible synaptic connections is incalculable i.e. approximately one million billion connections (Edelman, 2004). Recurrent neural architectures with their countless possibilities continuously emerge as functional (electrical and chemical) and flexible. Being dynamic and distributed because of their constant evolution and mutability, the question arises of the stability of traditional concepts like representation, image or motivation, still relevant in the field of tourism studies.

A fundamental issue to mention is consciousness or intentionality. In recent years the importance placed on different explications of consciousness has intensified. It is a field which is strictly connected with the tourist and with one of the salient points of the methodological debate in social sciences, where the converging ideas of people from different disciplines, even philosophy, like Searle, can be discussed.

Searle uses intentionality in the footsteps of Brentano and Husserl, but in a realistic perspective:

> On my view mental phenomena are biologically based: they are both caused by the operations of the brain and realized in the structure of the brain. On this view, consciousness and intentionality are as much a part of human biology as digestion or the circulation of the blood. [...] The correct solution to the 'mind-body problem' lies not in denying the reality of mental phenomena, but in properly appreciating their biological nature (Searle 1983: IX).

Also what is defined as the *Mystery of consciousness* is developed through the topic that consciousness is a natural, biological phenomenon, which does not pertain to the traditional categories of the mental or physical ones (Searle, 1997).

Edelman, the most prominent figure in neuroscience, distinguishes between primary and higher-order consciousness.

Primary consciousness is fundamental consciousness; higher-order conciousness is defined as 'The capability to be conscious of being conscious' (Edelman 2004, 161), and is present only in animals with semantic abilities like chimpanzees or linguistic abilities like humans, where the latter are also able to have a social concept of the self and concepts of the past and future.

Core issue of Edelman's perspective is the TNGS, that is the theory of neuronal group selection, consisting of three points

> (1) Developmental selection and (2) experiential selection, both of which operate on repertories of neural variants, and (3) Reentry, a key process assuring spatiotemporal correlation and conscious integration. It is a global brain theory explaining diversity and integration in the central nervous system (Edelman, 2004: 179).

Coupled with the process of reentry, memory and learning play a very relevant function, expounded especially in *The Remembered Present*: 'if no comparison took place between value and past categorization to form a special memory, consciousness would not appear' (Edelman, 1989: 102). The core issue is very simple, and is compatible with the most recent theories of memory, which have demolished the old myth of the memory as a storehouse. What we perceive is influenced by our individual past and present inputs.

One key concept in the philosophy of mind when considering whether consciousness exists is called qualia, which represent an important issue also in Edelman's theorization. He defines *quale; qualia* as a collection of personal and subjective experiences, feelings and sensations accompanying consciousness, stating that the hypothesis of *qualia* distinguishes higher-order consciousness (Edelman, 2004: 173). Certainly, compared with other radical neuroscientists even inclined to eliminate *qualia*, Edelman's attitude sounds quite balanced, but insufficient to prove the extent of the phenomenon (Oliverio, 1999). It is obvious that *qualia* are fundamental and constitutive aspects of human experience in general and of the tourist experience in particular.

As far as cognitivism is concerned, the main trend amongst opponents is to identify it with functionalism. Searle's radical criticism of cognitivism is often quoted as the *Chinese Room* argument (Searle,1984). Of course cognitive psychology cannot be reduced to functionalism and computationalism. At the end of the 80s' there were also relevant accents on embodiment as an exit of cognitive psychology,   the so-called embodied cognitive science, and applications to psychology of the theory of dynamic complexes as in the case of the well-known *Being there: putting brain, body, and world together again* (Clark, 1997).

As a common attitude of people siding against functionalism,   Searle, Lakoff, Johnson and others like them have pointed out that thought is not transcendent, but depends critically on the body and brain. Grounded on Johnson's *The Body in the Mind* (1987) for the embodiment of concepts, Lakoff's book *Woman, Fire and Dangerous Things* (1987) was published in the same year as Edelman's *Neural Darwinism* (1987), which was trying to cast the basis of a general brain theory, coping also with the problem of perceptive categorization without Edelman's knowing Lakoff's book. Probably Lakoff did not know Edelman's, either (Edelman, 1993: 385). In Edelman's next book *The Remembered Present* (1989) the brain theory was extended to perceptive experience, to the formation of concepts and to language. Looking back, in the neuroscientist's perspective, his books integrate the works of Langacker, Lakoff and Johnson because they offer fundamental biological support to many proposals of these authors on the importance of embodiment for grammar and cognition (Edelman 1993:389-390).

At this point the disciplinary and epistemological *vexata quaestio* of the relationship with psychology should be introduced, which can only be done with brief references.

In a sense William James was right when, in his *Principles of Psychology,* he affirmed that the neural basis for consciousness was the whole brain. His impossible dream that physiology, psychology and philosophy could be joined into a single discipline is also somehow linked to this fundamental issue (James, 1950).

Concerning tourism studies, probably all the three disciplines would have been necessary but, in short, psychology did not give a substantial contribution, especially considering the significance of the tourist experience (Pearce, 1982; Pearce and Stringer, 1991).
Maslow's classification of human needs has often been used (Maslow, 1954). If behaviorism did not have much to offer to research concerned with an experienced tourist, it was cognitivism, especially in its version of environmental psychology, which opened some new prospects.

Csikszentmihalyi's suggestions (1988) should moreover be mentioned, as his concept of 'flow' was acknowledged by several scholars (Franklin, 2003; Cohen, 2004), with some specific application especially to extreme adventure tourism.

Anyway, up to now no global framework has been individuated, and therefore the fields of the mind-body debate and neuroscience appear to be an indispensable point of reference.

A tangible scientific confirmation of neuroscience is given by so-called brain imaging, especially through PET, which provides information non invasively, and is beyond the realm of vague hypotheses. Functional brain imaging allows the individualization of areas of the brain which are more active than the surrounding ones. For example, the attempt to evoke a memory implies an increased activity of the frontal cortex, an emotion raises an increased activity of the limbic system etc. Yet it is still materially impossible to verify through brain imaging the tourist's mental processes before, during and after the trip, even though some experiments researching their neurophysiological basis could be carried out, probably similar to those employed in neurolinguistic tests.

To sum up, it is impossible to consider human beings as disembodied substances, the 'I' is not bodiless and mental processes, rationality and human action have to be conceived in a theoretical framework which takes into account our embodied nature (Di Francesco, 2002). Anyway, even if we are not disembodied 'thinking' substances, the problem of the total reduction of psychological phenomena to a neurobiological description still remains an open issue, but Edelman's neural Darwinism seems to be compatible with other possible more traditional studies.

A 'Universe of Consciousness (Edelman and Tononi 2000) glimmers, and the proposal of an epistemology grounded in biology looks convincing with its related possibilities of exploring also language and the self on a different basis.

The Tourist Subject, Happiness and Language

Mind and body have been somehow 'put together again' denying Humpty's irreversible fate. In fact, the tourist's cyberbody has consequently turned out to be more complex than any biological and electronic product. Technologies, tending to shift the weight to the sensorimotor mode, have paradoxically contributed to the body's fullness and saturation. Significant results have come from neuroscience and the mind-body debate and the emerging of a higher-order consciousness linked with the concept of self, that can be identified in our case with the tourist subject.

At this point, in the light of the tourist's global evolutionary experience, some words can be spent on the tourist's collective 'imaginary'. This has been evolving from day to day through traditional and new potentials of the human mind, like fantasy, dreams, imagery, literature, movies and VR, in an increasingly complex development. The possibility for realizing imaginary phantasies which in traditional societies was reserved to play, feasts and rituals, can now be individuated in modern tourism, maintaining an intrinsic link with the body (Hennig, 1999: 101).

In a mainly cultural anthropological dimension, two poles of this 'imaginary' can be singled out, both imbued with myth, but differently articulated, one being the 'last island', while the other is about wellness, corresponding to different forms of tourism typologies but also interplaying.

On the more clearly determined pole of tourism, purity and primitivity emerge especially in the image of the last island. Strictly connected with these issues is often the dimension of liminality and the more or less carnivalesque 'communitas'.

The environment and the consequences of the above can be different, but the accent on the Search of the Earthly Paradise (Eliade, 1971) does not change, apparently steadfast even for scholars less open to this viewpoint. Among recent examples, Cohen deals with *ThePacific Islands from Utopian Myth to Consumer Product: The Disenchantment of Paradise* (Cohen, 2004:.255-56), while Sheller researches on *Demobilizing and remobilizing Caribbean Paradise* (Sheller, 2004) in the framework of a perspective of mobility, inglobing the one of tourism studies (Sheller and Urry 2006).

An associated aspect to be taken into consideration is the one of 'enchantments and disenchantments': paradoxically, 're-enchantment' is now being carried out by the industry not only through feelings, but also through more rational processes, even using an alleged scientific cognitive attitude (Selwyn, 2005).

On the opposite pole of the island, more akin to leisure, but still somehow bound with the myth of eternal youth, the simpler dimension of wellness can be pointed out, also fostered in varying degrees by the industry (Becker and Brittner, 2003). In this growing tourist sector re-enchantment can assume mixed forms of everyday life-leisure-tourism. The heritage of the Western spa tradition, fitness, but also Asian religions and esoteric new-age philosophies merge in an appealing cauldron, with a strong focus on polisensorial experience.

'Modern myths usually remain limited to symbolic level '(Hennig, 2002: 184). From an epistemological point of view, it is a discourse with a blurred subject, featured by the progressive erosion not of the myth itself in its different versions but of the myth's purity, and by the unavoidable absence of the diachronical dimension. In fact, in the tourist's perspective, myth is much more cyberintertwined, and going back to a preceding comparison, the mythical image of the 18th century male traveller is certainly totally different from that of a 21st century female space traveller.

Coupled with myth, but somehow behind and beyond myth, promising openings concerning the presence of the conscious subject come from the comprehensive concept of happiness, working both at a collective and individual level. Actually, although Enzensberger (1996) defines tourism as a Pyrrhic victory, the happiness factor is highlighted under the aspect of the implicit-explicit pursuit of happiness.

> The force, however, that drives vacationers to the beaches of their little holiday happiness is immense. It is the force of a blind and inarticulate rebellion, doomed to go under in the surf of its dialectics [..]. Tourism is based on the desire for the pursuit of happiness […].The images of happiness created by Romanticism will endure […] as long as we are unable to confront them with our own images'
> (Enzensberger, 1996:135).

This approach can turn out to be productive if centered on the tourist, in a more sociopsychological dimension. It is actually common talk to couple the tourist with the pursuit of happiness, the consciousness of happiness, the conspicuous experience of it.

At this point the way is open to interdisciplinary approaches, which have become rather common between neurobiology, psychology, economics and linguistics.

What we would like to explore is the recent encounter between the concepts of happiness, economics, cognitive psychology (Kahneman, 2005; Kahneman and Krueger, 2006) and the possible dimension of language, to be caught through the narrative of tourism (Bruner 2005).

For many years the Nobel prize winner Kahneman went ahead with experiments together with the psychologist Tversky, starting from the problem of economic choices. The link between economics and happiness commonly held-up to the 19th century, is not at all surprising. It recently gave rise to a promising line of research (Layard, 2005; Bruni and Porta 2005).

The concept of happiness is linked with consciousness, with the self, that is the conscious subject, and with language in general. Above all it is a matter of differentiating between people's two experiences of happiness, that is, subjective and objective, and results have also revealed surprising gaps between various countries (Kahneman 2005, Kahneman and Krueger, 2006).

Moreover, in a study on holidays, substantial discrepancies between the memory of pleasure felt when on holiday and the actual experience of pleasure were found (Kahneman 2005).

The *experiencing self* does not count, but the *remembering self's* view does, the self that records and pigeonholes experiences. Again according to Kahneman, various discoveries have been made:

> The first is that changes in state produce subjective values, not the states themselves. The second is that only the 'peaks' (positive or negative) count, and how the episodes end, not how long they actually last, nor the periods of normality [...]. Actually I am in favour of a hybrid model, of a more complex and integrated vision, which takes into consideration both the experience itself and the way it is pigeonholed, classified, lived over again and recounted. It is important for us to remember, live over again and give an account of our experiences, even if the original experience was different for us. A psychological and economic study of happiness, of wellness, cannot neglect this aspect (Piattelli Palmarini, 2006).

It is now the moment to link up the psychological and the linguistic dimensions. This perspective is not only compatible with the neurophysiological basis of the tourist's consciousness, but expands the view of neuroscience to the field of memory and covers the broad space which neuroscience left open to cognitive psychology and its embodied categories. Attention should be given to the tourist's language, as a component of her/his 'imaginary'.

In his sociolinguistic view, tourism had been seen by Dann (1996) as a language of control, but already in 1996 the possibility was envisaged for tourists, in turn, to contribute further to this language by communicating their experiences; nowadays, as we have stated, the autonomy of the tourist emerges, with the accent on a more subjective language controlled by the tourist herself/himself (Dann, 2005).

Interesting suggestions come from Bruner (2005), who distinguishes different stages of narrative related to the tourist experience and also mentions Kahneman. In our case it is not the global narrative encompassing much broader aspects related to the whole complexity of the tourism phenomenon including destinations, interactions, itineraries, but just a very concentrated moment of the tourist's narrative, the narrative of her/his embodied experience, **somehow related to the** epiphanic sense of the tourist experience suggested by Hom Cary, who also distinguishes between experience and representation (2004). As Bruner puts it 'After the trip, narratives of travel are put in order, consolidated, condensed, and made more coherent' (Bruner, 2005: 17). Memory is rightly seen as a proper construction, according to recent researches on memory, considering the stage of codification and of building up memories (Schacter, 2001).

This 'narrative of happiness' leads us for the moment to focus on some aspects, like the individual and social dimension of language, the primacy of semantics (entire worlds are sometimes associated with a single word, e.g. the name of a tourist destination) and memory's pre-eminent function. Clearly all the different functions of language and its connections with memory should be given further study.

In his fruitful interdisciplinary research Kahneman states that there is a radical difference between the evaluation of on-the-spot, immediate happiness and overall happiness on reflection, implicitly reminding us of the psycho-environmental scheme of the tourist travel experience and of the post-trip moment of recollection (Liebman, 1993).

Subjective happiness, often called 'indescribable' in everyday life, comes true in this perspective only if it is brought to the mind and described.

The tourist's whole psychic life participates in a simple memory. The tourist pulls the threads of his narrative tissue, not as in Dennett's proposal, where   consciousness is more or less a continuing flow of drafts of texts, and the texts are constantly being edited and re-edited, implying the disappearance of the subject (Di Francesco, 2002). On the contrary, at the moment of memory's codification, the tourist is to be considered as a sovereign subject, lord of his life, consolidating her/his autobiography, coupling the embodied self with the dimension of memories (Schacter, 2001).

Being a tourist would then identify with enforcing her/his tourist experience through memory and language.

Conclusion

Coming back more precisely to the tourist experience, it seems to us that the two apparently distant perspectives of Cohen and Kahneman converge, although both start from completely different fields. In controversy with MacCannell, Cohen argues that otherness and sameness, like authenticity, '[...] are not dichotomous concepts but rather end-points of a continuum' (Cohen, 2004: 318). Moreover they are not absolute concepts, but context-dependent and socially constructed, in the framework of the progressive de-differentiation in favor of leisure.

Kahneman states – almost by chance - that also the period of time and normality can be the cause of happiness:

It would be senseless to want to give the same weigth to every moment of the same life subsequently, but it would be just as senseless to take seriously the idea that 'one day as a lion is better than a hundred days as a sheep' (Piattelli Palmarini, 2006).

Let us go back to the subject who weaves the threads of his own memory *ex post,* who conceptualizes his experience, puts it into words and makes a narrative of it. The tissue is subjective but embodied, it is a moment of autopoiesis where 'the Brain – is wider than the Sky –' (Dickinson, [c.1862] 2004; Edelman, 2004: VII).

Moreover, it teaches us that not only peaks and great distances exist; there is also the sublime banality of life, with the tourist enhancing it all.

References

Appadurai, A. (2001) Modernità in polvere. Dimensioni culturali della globalizzazione. Roma: Meltemi.

Baudrillard, J. (1983) Simulations. New York: Semiotext.

Bauman, Z. (1998) Globalization. The Human Consequences. London: Polity Press.

Becker, C. and Brittner, A. (2003) 'Wellness-Tourismus in Deutschland und den USA. Ein Vergleich', Voyage – Jahrbuch für Reise- & Tourismusforschung . Schwerpunkt Körper auf Reisen 6: 81-89.

Bloch, E. (1959) Das Prinzip Hoffnung. Frankfurt a.M.: suhrkamp.

Boorstin, D. (1964) The Image: A Guide to Pseudo Events in America. Nuw York: Harper & Row.

Bruner, E. (2005) The role of Narrative in Tourism. Paper presented to the Berkeley conference 'On Voyage. New Directions in Tourism Theory', October 2005.

Bruni, L. and Porta P.L. (2005) Economics and Happiness. Framing the Analysis. Oxford: Oxford University Press.

Clark, A. (1997), Being There: Putting Brain, Body, and World Together Again. Cambridge Ma.: The MIT Press.

Cohen, E. (1979), 'A Phenomenology of Tourist Experiences', Sociology: The Journal of the British Sociological Association, 13 (2): 179-201.

—. (1988) 'Authenticity and Commoditization in Tourism', Annals of Tourism Research 15 (3): 371-386.

—. (2000a), 'Experience', pp. 215-216 in Jafari, J. (ed.) Encyclopedia of Tourism. London: Routledge.

—. (2000b), 'Sociology', pp.544-547 in Jafari, J. (ed.), Encyclopedia of Tourism. London: Routledge.

—. (2004) Contemporary Tourism. Diversity and Change. Amsterdam: Elsevier.

Csikszentmihalyi, M. (1988) Beyond Boredom and Anxiety. San Francisco: Jossey-Bass.

Damasio, A. 1994 Descartes' Error. Emotions, Reason and the Human Brain. New York: Putnam.

—. (2003) Looking for Spinoza. Orlando, Florida: Harcourt.

Dann, G.M.S. (1996) The Language of Tourism. A Sociolinguistic Perspective. Wallingford: CAB International.

—. (2005) Remodelling the Language of Tourism. From Monologue to Dialogue and Trialogue. Paper presented to the International Academy for the Study of Tourism Beijing, July 2005.

Dann, G. and Jacobsen, J. (2002) 'Leading the Tourist by the Nose', pp. 209-235 in Dann, G.M.S. (ed.) The Tourist as a Metaphor of the Social World. Wallingford: CABI.

Dann, G.M.S. and Liebman Parrinello, G. (2005) From Travelogue to Travelblog: (Re)-negotiating Tourist Identity. Paper presented to the Berkeley conference 'On Voyage. New Directions in Tourism Theory', October 2005.

de Kerckhove, D. (2005) 'An interview with Derrick de Kerckhove by Alvaro Bermejo', Communication in evolution: Social and technological transformation. The McLuhan program in culture and technology (consulted July 2005): http://www.mcluhan.utoronto.ca/article_communicationevolution.htm.

Dennett, D.C. (1991) Consciousness Explained. Boston: Little. Brown & Co.

Dickinson, E. (2004 [1862 c.] Wider Than the Sky, quoted by Edelman, G.M. (2004) Wider than the Sky. New Haven and London: Yale University Press.

Di Francesco, M. (2002) Introduzione alla filosofia della mente. Roma: Carocci.

Eco, U. (1987) Faith in Fakes. London: Secker and Warburg.

Edelman, G.M. (1987) Neural Darwinism. New York: Basic Books.

—. (1989) The Remembered Present. New York: Basic Books.

—. (1993) Sulla materia della mente (Bright Air, Brilliant Fire). Milan: Adelphi.

—. (2004) Wider than the Sky. New Haven and London: Yale University Press.

Edelman, G.M. and Tononi, G. (2000) A Universe of Consciousness. New York: Basic Books.

Eliade, M. (1971) The Myth of Eternal Return. Princeton: Princeton University Press.

Enzensberger, H.M. (1996 [1958]) 'A Theory of Tourism', New German Critique 68: 117-135.

Franklin, A. (2003) Tourism. An Introduction. London: Sage.

Gehlen, A. (1984) L'uomo nell'era della tecnica. Milano: SugarCo Edizioni.

—. (1993) Der Mensch. Seine Natur und seine Stellung in der Welt. Frankfurt a.M.: V. Klostermann.

Graburn, N. (1983) 'The Anthropology of Tourism', Annals of Tourism Research – Special Issue 10 (1):9-33.

—. (1995) 'The Past in the Present in Japan: Nostalgia and Neotraditionalism in Contemporary Japanese Domestic Tourism', pp.47-70 in R.W. Butler and D. Pearce (eds) Changes in Tourism. People, Places, Processes. London: Routledge.

Graburn, N. with Diane Barthel-Bouchet (2001), 'Relocating the Tourist', International Sociology 16 (2): 147-158.

Hennig, C. (1997) 'Jenseits des Alltags', Voyage- Jahrbuch für Reise & Tourismusforschung' 1: 35-60.

—. (1999) Reiselust. Frankfurt.a.M.: suhrkamp.

—. (2002) 'Tourism: Enacting Modern Myths', pp.169-187 in Dann, G.M.S. (ed.) The Tourist as a Metaphor of the Social World. Wallingford: CABI.

Hom Cary, S. (2004) 'The Tourist Moment', Annals of Tourism Research 31 (1): 61-77.

Hunziker, W. and Krapf, K. (1942) Grundriß der allgemeinen Fremdenverkehrslehre. Zürich: Polygraphischer Verlag AG.

James, W. (1950) The Principles of Psychology. New York: Dover.

Johnson, M. (1987) The Body in the Mind: The Bodily Basis of Meaning, Imagination, and Reason. Chicago: Chicago University Press.

Johnston, L. (2001) '(Other) Bodies and Tourism Studies', Annals of Tourism Research 29 (1): 180-201

Kahneman, D. (2005) 'Il Pil dei momenti felici', Il Sole – 24 ore, 4/3 2005: 31.

Kahneman, D. and Krueger, A.B. (2006) 'Developments in the Measurement of Subjective Well-Being', Journal of Economic Perspectives 20 (1): 3-24.

Kern, S. (1988) Il tempo e lo spazio: La percezione del mondo fra Otto e Novecento, Bologna: Il Mulino.

Lanfant, M.-F. (1995) 'International Tourism, Internationalization and the Challenge to Identity', pp. 24-43   in M.-F.Lanfant et al. (eds) International Tourism. Identity and Change. London: Sage.

Lakoff, G. (1987) Women, Fire, and Dangerous Things. Chicago and London: The University of Chicago Press.

Lakoff, G. and Johnson, M. (1980) Metaphors We Live By. Chicago: Chicago University Press.

Lash, S. and Urry, J. (1994) The Economies of Signs and Space. London: Sage.

Layard, R. (2005) Happiness: Lessons from a New Science. London: Penguin.

Lévy, P. (1996) L'intelligenza collettiva. Milan: Feltrinelli.

—. (2000) Cybercultura. Gli usi sociali delle nuove tecnologie. Milan: Feltrinelli.

Liebman Parrinello, G. (1993) 'Motivation and Anticipation in Post-industrial Tourism', Annals of Tourism Research 20 (2): 233-249.

—. (2001) 'The Technological Body in Tourism Research and Praxis', International Sociology 16 (2): 205-219.

MacCannell, D. (1973) 'Staged Authenticity; Arrangements of Social Space in Tourist Settings', American Journal of Sociology 79 (3): 589-603.

—. (1976) The Tourist: A New Theory of the Leisure Class. New York: Schocken Books.

Maldonado, T. (1989) Il futuro della modernità. Milan: Feltrinelli.

—. (1997) Critica della ragione informatica. Milan: Feltrinelli.

Maslow, A.H. (1954) Motivation and Personality. New York: Harper & Brothers.

McIntosh, W, and Goeldner, C.R. (1995) Tourism Principles, Practices and Philosophies. New York: Wiley.

McLuhan, M. (1967) Understanding Media. The Extensions of Man. London: Sphere Books.

Meetham, K. (2001) Tourism in Global Society Place Culture Consumption. Houndmills: Palgrave.

Minca, C. (1998) Spazi effimeri. Padua: CEDAM.

Morkham, B, and Staiff, R. (2002) 'Perception and Subjectivity', pp. 297-316 in G.M.S. Dann (ed.) The Tourist as a Metaphor of the Social World. Wallingford: CABI.

Oliverio, A. (1999) Esplorare la mente: Il cervello fra filosofia e biologia. Milan: Raffaello Cortina Editore.

Olsen, K. (2002) 'Authenticity as a Concept in Tourism Research', Tourist Studies 2: 159-81

Pearce, P.L. (1982) The Social Psychology of Tourist Behaviour. Oxford: Pergamon Press.

—. (1988) The Ulysses Factor: Evaluating Visitors in Tourist Settings. New York: Springer Verlag.

Pearce, P.L. and Stringer, P.F. (1991) 'Psychology and Tourism', Annals of Tourism Research 18 (1): 136-145.

Perkins, H.C. and Thorns, D.C. (2001) 'Gazing or Performing? Reflections on Urry's Tourist Gaze in the Context of Contemporary Experience in the Antipodes', International Sociology 16 (2) 185-204.

Piattelli Palmarini, M. (2006) 'Intervista al premio Nobel che ha elaborato con lo psicologo David Schkade la teoria della 'focusing illusion', Corriere della Sera 10/29: 31.

Poon, A. (1993) Tourism, technologies and competitive strategies. Wallingford: CABI.

Prideaux, B. (2002) 'The Cybertourist', pp. 317-39 in G.M.S. Dann (ed.) The Tourist as a Metaphor of the Social World. Wallingford: CABI.

Quan, S. and Wang, N. (2004) 'Towards a Structural Model of the Tourist Experience: an Illustration from Food Experiences in Tourism', Tourism Management 25: 297-305.

Rojek, C. and Urry, J. (1997) 'Transformations of Travel and Theory' in C. Rojek and J. Urry, (eds.) Touring Cultures: Transformations of Travel and Theory. London: Routledge.

Ryan, C. (2002) (ed.) The Tourist Experience. London: Continuum.

Schacter, D. Alla ricerca della memoria (Searching for Memory). Turin: Einaudi.

Schivelbusch, W. (1977) Geschichte der Eisenbahnreise. Münich: Hanser.

Searle, J. (1983) Intentionality: An Essay in the Philosophy of Mind. Cambridge Ma.: The MIT Press.

—. (1984) Minds, Brains and Science. Cambridge Ma: Harvard University Press.

Ryan, C. (2002) (ed.) The Tourist Experience. London: Continuum.

Schacter, D. Alla ricerca della memoria (Searching for Memory). Turin: Einaudi.

Schivelbusch, W. (1977) Geschichte der Eisenbahnreise. Münich: Hanser.

Searle, J. (1983) Intentionality: An Essay in the Philosophy of Mind. Cambridge Ma.: The MIT Press.

—. (1984) Minds, Brains and Science. Cambridge Ma: Harvard University Press.

—. (1997) The Mystery of Consciousness. London: Granta Books.

Selwyn, T. (2000) 'Sun, Sand, Sea and Sex', p.564 in Jafari, J. (ed.), Encyclopedia of Tourism. London: Routledge.

—. (2005) The Political-Economy of Enchantment: Formations in the Anthropology of Tourism. Paper presented to the Wageningen Conference of the ISA RC 50, June 2005.

Sheldon, P. (1998) Tourism and Information Technology. Wallingford: CAB International.

Sheller, M. (2004) 'Demobilizing and Remobilizing Caribbean Paradise', pp. 13-21 in Sheller, M. and Urry, J.(eds) Tourism Mobilities. London: Routledge.

Sheller, M. and Urry, J.(eds) (2004) Tourism Mobilities. London: Routledge.

Smith, V.L. (2001) 'Space Tourism', Annals of Tourism Research 28 (1): 238-40.

Swain, M. (1995) 'Gender in Tourism', Annals of Tourism Research 22: 247-66.

Theobald, W. F. (ed.) (2005) Global Tourism. Third edition. Elsevier: Amsterdam.

Urbain, J.D. (1994) Sur la plage. Paris: Payot.

Uriely, N. (2005) 'The Tourist Experience. Conceptual Developments', Annals of Tourism Research 32 (1): 199-216.

Urry, J. (1990) The Tourist Gaze. London: Sage.

—. (2001) 'The Tourist Gaze and Beyond. An Interview with John Urry', Tourist Studies 1(2): 115-131.

Varela, F. et al. (1996) The Embodied Mind. Cognitive Science and Human Experience. Cambridge Ma: The MIT Press.

Veijola, S. and Jokinen, E. (1994) 'The Body in Tourism', Theory, Culture & Society 11 (3):125-151.

Wang, N. (1999) 'Rethinking Authenticiy in Tourism Experience', Annals of Tourism Research 26 (2): 349-370.

—. (2000) Tourism and Modernity: a Sociological Analysis. Oxford: Elsevier.

—. (2003) 'Tourismus und Körper. Soziologische Betrachtungen', Voyage -Jahrbuch für Reise- & Tourismusforschung. Schwerpunktthema: Körper auf Reisen 6: 127-134.

Williams, A.P. and Hobson, J.S.P. (1995) 'Virtual Reality and Tourism. Fact or Fantasy?', Tourism Management 16 (6): 429-35.

DANCING BODIES AND CYBER BODIES: FROM BAROQUE TO ULTRAMODERNITY IN BRAZILIAN FAVELAS

ROBERTO MOTTA

Abstract: The bodies and their dancing have undergone a change of great magnitude in the Brazilian favelas. In the beginning, thought was encapsulated in the motions of dancing. Conceptual systems were implied in gestures. The motions of the body were not subject to time as a measure of productivity, but followed their own rhythm. This changes, however, with the growth of the market as the measure of everything and with the process of acculturation, which applies the rationalism of Western culture to all cognitive systems. The spontaneity of the body recedes. This can be specially observed in the so-called Afro-Brazilian religions. Candomblé is increasingly replaced by Umbanda or by new forms of syncretism based on the theories of anthropologists and sociologists. Bodily motions are more and more reduced to icons or signs. And finally it is a whole problem of historical interpretation that is raised. Is the body irretrievably lost or can it survive among some ethnic and cultural nuclei? Is there a return to the body in the post-modern age or is the thought of such return equivalent to the utopia of a return to the golden age?

Preliminary Remarks

This paper[1] has a thesis which will hopefully be demonstrated through the inspection of the evidence furnished by the Afro-Brazilian religions[2] and their evolution during the 20th century. This thesis, as the author is fully aware, represents a commonplace in the social sciences, which he intends to rehearse in this paper, all the more so as he believes that it constitutes

---

[1] The author wishes to express his gratitude to Ivan Varga and Bianca Maria Pirani, editors of this volume, for their pertinent queries and suggestions, which were very much taken into account by the author in the preparation of the final version of this paper.

[2] Mainly those situated in the area of Recife, in Northeastern Brazil, where the author has done extensive field work.

one of the solidest acquisitions of sociological and anthropological theory, apparently endangered, nowadays, by the rise of interpretations that are far less well founded on the available historical evidence. The thesis –or the congery of theses—upheld here is very simple. The body recedes[3] with modernization.[4] Now, modernization may be defined in an x number of ways. It may even so happen that, according to some of those ways, the process of modernization is now closed. Thus I do not intend to deny[5] the pertinence of Jean-François Lyotard's (1979) analysis concerning 'the end of grand narratives'[6], that is the collapse of those theories, like Hegel's, Marx's and others, which imply, as working within history, a tendency that guides it toward progress. There have of course been, in the last decades, wide changes in fashions, styles, sexual orientations, artistic schools, literary trends, philosophical and social theories, religious expressions, and so on and so forth.[7] Nothing precludes us from calling

---

[3] This should be understood with 'esprit de finesse', that is, with all the subleties and distinctions the subject demands. In strict terms, the body itself, with its urges and its drives, obviously does not disappear nor recedes. But its spontaneity does. This is of course relative. No culture has been able to eliminate the force of the body. This force, nevertheless, as it is all too clear in our time, is subordinated to the requirements of the process of accumulation and to the need for order and rationality, which are the hallmarks of modernity understood according to Marxian and Weberian principles. Thus, to give again a trivial example, the sex drive, with all its strength, has not been eliminated. Yet sexual practices are subject to a process of rationalization (or regulation) that leads, among other things, to the low birth rates prevalent in Western countries.

[4] Regarding the status of the body in modernity, an indeed extensive literature is available. Beside Tönnies (1988) and Weber (1958, 1978), the writings of Norbert Elias loom large in this context (especially Elias, 1978a and 1978b). The work of Michel Foucault (among others Foucault, 1984) also bears strongly on our subject. Norbert Elias was the object of a detailed attempt at refutation by Hans Peter Duerr in *Nudité & Pudeur: Le Mythe du Processus de Civilisation*, (Duerr, 1998). But all Duerr seems to have been able to demonstrate was that the repression of the body's spontaneity occurs also outside modern Western society, indeed in primitive society, but under special circumstances and without characterizing the ethos, the whole cultural process of those societies. Neither as used by Max Weber nor by the present commentator the idea of a 'spontaneous, impulsive enjoyment' of the tendencies of the body implies the absence of cultural regulation even of this spontaneous and impulsive enjoyment. If we wish, we can perfectly say that both spontaneity and rationalization (with the repression it entails) are both products of culture.

[5] Nor, for that matter, to endorse.

[6] La fin des grands récits

[7] The year 1968 may perhaps be seen a milestone, with the students's revolts in

'postmodern' the end results of these changes. Yet, it is argued in this paper that two of the basic characteristics of modernity have come to stay, or at least, have not been changed and give no signs of receding. I refer, on one hand, along lines suggested by Karl Marx, to the rise of Capitalism with its corollaries of an accumulation that always tends to increase, under the direction of a spirit of calculation, which is very much what Marx, right at the beginning of *Capital*, calls abstraction, and the commodification of very nearly every social or cultural manifestation. On the other hand, I refer to the rise of rationalization such as it is presented, in ideal-typical fashion, in Max Weber's writings.

Rationalization has as its 'most urgent task the destruction of spontaneous, impulsive enjoyment' (Weber, 1958: 119), associated with the tendencies of the body, that is, with 'the actor's specific affects and feeling states [as they tend to satisfy] a need for revenge, sensual gratification, devotion, contemplative bliss, or for working emotional tensions' (Weber, 1978:25). Max Weber is, par excellence, the theoretician of rationalization[8]. This concept is akin to Marx's abstraction.[9] Yet Max Weber denies the primacy Marx attributes, in the process of historical development, to the economic basis (infrastructure) of society. Weber replaces the notion of such primacy by the concept of an elective affinity between, on one hand, values and attitudes stemming from religious beliefs and commitments and, on the other hand, the characteristics of the infrastructure.

Marx and Weber are treated as the authors of some of the major analyses of the basic tendencies of our time. To say it with the words of a German commentator, they are situated right at the summit of history, 'am Spitze der Geschichte'[10] (Sombart, 1958:43), and this not on account of

---

Paris, New York and elsewhere, the acceleration of the movement of sexual liberation and other trends. Let us also remember the religious transformations associated, in the Roman Catholic Church, with the II Vatican Council (1962-1965).

[8] This, however, by no means is the same as saying that only Max Weber dealt with rationality. Even apart from earlier forms of this concept in the writings of Hegel and of Marx, it also plays leading roles, to give but two eminent examples, in Ferdinand Tönnies (1978) and Werner Sombart (1913).

[9] The opposition concrete/abstract plays a major role right in the first part of volume I of Capital, 'Commodities and Money', from which I draw the following quotation: 'There is nothing left [in products] but what is common to them all; all are reduced to one and the same sort of labour, human labour in the abstract (Marx, 1967:38).

[10] In this passage Nicolaus Sombart alludes primarily to the early founders of Sociology, but are not Marx and Weber also founders of Sociology?

their metaphysical ideas about the meaning of History, but rather because, due to the exactitude of their interpretation, they, to use now the words of yet another German commentator, turn into 'the clearest consciousness of our time by itself'[11] (Karl Jaspers, as quoted by Abramowski, 1966: 9).[12]

The body, or the spontaneity of the body, recedes therefore with the process of expansion of modern Capitalism accompanied by the process of rationalization with which it is inextricably bound. Some revolutions have receded. This is apparently the case of the Communist revolution,[13] at least in Central and Eastern Europe. This is the great event symbolized by the fall of the Berlin Wall in 1989. Yet (giving this expression a meaning similar to that which it has in Berger, 1986) the 'capitalist revolution' has not receded. Now, if the recession of the spontaneity of the body is a consequence of the rise of Capitalism, with its emphasis on the ever growing production of commodities and its all-encompassing process of rationalization and if Capitalism has not receded, it seems likely that a return to a 'civilization of the body', that is, to a culture in which the body keeps its spontaneity and impulsive enjoyment, is a utopian dream, as the iron cage of Capitalism and rationality is by no means broken.

Part and parcel of the central thesis of this paper is that Brazil, with its 'Afro-Brazilian cults', has also been caught in that cage. This does not mean that the country and its religions have lost their specific cultural characteristics. Indeed, one of the questions raised by the expansion of the

---

[11] Jasper's reference is directed primarily to Weber. In his German original he says 'To many of us Max Weber appears above all as a philosopher. [...] A philosopher is more than one who simply knows. [...] He is a representative of the time. [...] He is the living heart of the time, but he is is even more than this. He gives the time the capacity of expressing itself, he is the mirror in which it recognizes itself and thus he enables the time to determine its direction' (Jaspers, 1988:32-36). This, by the way, is very much the equivalent of attributing to Weber, and, in an analogous way, to Marx (although the latter is not mentioned by Jaspers) the honor of being the outstanding representatives of Hegel's Absolute Spirit in their periods.

[12] A major French interpreter, Jean Hyppolite (1983:20), says that 'Hegel discovers, beyond the morality (*Moralität*) [...] the living reality of customs and institutions (*Sittlichkeit*)' Hypolitte also says that 'what concerns Hegel is to discover the spirit of a religion, the spirit of a people, he wants to forge new concepts capable of expressing man's historical life, his existence within a people and a history' Ibid.:13). The same, with but small change, applies equally well to Marx and Weber. Now, inasmuch as Anthropology is the science of 'customs and institutions' it might even be claimed that Marx and Weber were not only eminent sociologists but also eminent anthropologists.

[13] This is represented by the end of both 'real-existing socialism' and the so-called 'people's democracies ' and the 'dictatorships of the proletariat'.

capitalist revolution concerns its capacity to integrate all other cultures and societies into its framework of productiveness and rationalization. Let us avoid the temptation of an all too easy pseudo-egalitarianism in believing that all peoples of all cultures can assimilate out of hand the basic attitudes of Capitalism, or, if we prefer, modernity. To this question there are no cogent, purely theoretical solutions. It can only be answered by long range historical observation,[14] lasting at times for generations. But whether or not the whole world has been successfully modernized, it is in any event true, as it has been known for a long time, that the iron cage of rationality, that penetrates and permeates every aspect of the society, the culture and the economy of the leading countries of the Western world, has also influenced, challenged and changed the rest of the world. This of course does not imply the successful westernization or modernization of the whole world, but it does imply that no country failed to be affected by the diffusion of rationality.

The Body Changes in Brazil

The second part of this paper is devoted to a punctual issue, with the goal of buttressing the central thesis of this paper concerning the permanence of commodification, abstraction and rationality. By using the metaphors of the dancing body and of the cyber body, I intend to demonstrate how, under the influence of the capitalist revolution, which for a long time has been a globalized phenomenon, the conception of body and, indeed, the time of the body, have changed in Brazil. However, rather than studying the role and the interpretation of the body in Brazilian culture as a whole, I will focus on a very specific topic, which has been my main area of field work. I refer to the so-called Afro-Brazilian religions. As I interpret them, they have originated not only from the conservation, in Brazil, of an African memory, but have also resulted from a process of syncretism with the baroque Catholicism of the colonial period. This kind of Catholicism stressed the image rather than abstract thought and was, therefore, uncongenial to the development of a modern Capitalism with its requirements of abstraction and rationalization.[15]

---

[14] Thus the recent Indian economic boom –or perhaps even the earlier development of the 'Asian Tigers'-- may have rendered outdated many of the conclusions of Gunnar Myrdal's essay An Asian Drama (1963).

[15] This is stressed by many students of baroque religious art. In his outstanding Le Baroque Victor-Lucien Tapié stresses that Baroque art 'provides concrete and simple witness of a familiar protection acting in everyday life, which endows some statues with a similarly concrete personality thus encouraging the faithful to

The African religions were also religions of the image. The greatest
affinity between popular Baroque Catholicism and the **Candomblé**, seen
as the prototype of the syncretistic Afro-Brazilian religions, lies in the
markedly concrete character of their forms of devotion, very often
addressed to saints that at least theoretically were seen not as gods in their
own right but rather as saints or intercessors between the faithful and the
Supreme Deity. In agreement with the basic concreteness in the
conception of the supernatural forces, there is also a marked prevalence of
gifts and offerings in the relationship of the faithful to the saints, who are
believed to reciprocate in terms of patronage and protection.[16] Gifts can be
of many kinds. They may consist in vocal prayers, in relatively small
objects such as candles or flowers, in feasts, pilgrimages, or the
performance of rare and difficult actions. The gift may also consist in the
offer of money to the saints, or to their priests, temples, shrines, convents,
etc. And it may also take the form of animal slaughter, which was widely
practiced in the Greco-Roman world but was foreign to the Ibero-
American religious tradition. It was nevertheless introduced –having
survived until the 21st century-- into certain areas of Latin America
(Brazil, Cuba), and grafted into the pattern of popular devotion through the
influence of the religions of West Africa.

Thus, the religion that prevailed in traditional Brazil and which, in
varying degrees of syncretism with the African-influenced cults, is still
powerful nowadays, was a kind of sacrificial religion (even when blood
sacrifice was absent) in that it was very much oriented to the 'dyadic
contract' between men and saints. As it is not difficult to imagine (with a
little help from Max Weber), this kind of religion, due to the lack of
'elective affinities' with such process, was hardly conducive to a
generalized process of rationalization of people's behavior and society.

In contrast to the sacrificial religions, in ethical religions, the object of
the gift becomes the very person of the believer, with its interiority and
freedom. Performances become secondary and, in strict terms, irrelevant.
The faithful conceives, at least in ideal-typical terms, of no other form of

---

have a special trust in them, to visit them in collective or individual pilgrimage,
treating them as living and powerful beings [...] the risk of idolatry that those
practices entailed was fully realized [by the Church] but it was held that an
abstract form of worship would be conducive to [religious] indifference (Tapié,
2002: 29). This passage summarizes a basic trait of both Counter-Reformation
popular Catholicism and of the syncretic Afro-Brazilian cults.
[16] This corresponds to the 'dyadic contract' studied by Latin-Americans,
outstanding among them Foster (1967), whose conclusions are by no means
outdated.

sacrifice but that of his own person and his own freedom, his behavior counting but as a mirror of his options. Let us add, however, some qualifications to this perhaps excessively clear-cut dichotomy. Max Weber himself, in spite of his copious use of that opposition in *The Protestant Ethic and the Spirit of Capitalism*, sounds a critical note, when, in his lengthy chapter of 'The Sociology of Religion' of *Economy and Society* (Weber 1978), he points out that the ethical religions, a hallmark of which is a rationalized 'conduct of life', may themselves have 'magical origins', stemming from 'the rationalization of taboo'. As he puts it, in a spirit certainly quite different from that of *The Protestant Ethic*:

> The rationalization of taboos leads ultimately to a system of norms according to which certain actions are permanently constructed as religious abominations subject to sanctions. [...] In this manner there arises an ethical system, the ultimate warrant of which is taboo (Weber 1978:433).[17]

Gilberto Freyre, who is considered as a leading interpreter of Brazilian culture, furnishes a very apt description of the festive religion once prevalent in Brazil. According to him

> What took place in our country was a deep-going and fraternal association of values and sentiments. [...] This was a kind of fraternization that could only with difficulty have been realized under any other type of Christianity than that which dominated Brazil during its formative period; a more clerical type, more ascetic, more orthodox, Calvinistic or strictly Catholic, would by no means have been so favorable as that mild brand of household religion, emanating from the chapels of the Big Houses, which presided over the development of Brazilian society, with family relations, one might say, prevailing between the saints and man and with churches that were always having feasts, baptisms, marriages, with banners, saints, chrisms, and novenas. It was this domestic, lyric, and festive Christianity, with its humanly friendly male and female saints and its Our Ladies as godmothers to the young, that created the first spiritual, moral, and aesthetic bonds between the Negroes and the Brazilian family and its culture. [...] The freedom of the slave to preserve, and even to display in public --at first on Epiphany Eve, and later on Christmas Night, New Year's Night, and during the three days of Carnival¾ the forms and accessories of his mystical, fetishistic, and totemistic culture affords a very good idea of the

---

[17] Rudolf Bultmann (1984 new edition) suggests a somewhat intermediary category between the sacrificial (or cultic) and the ethical religions. Thus he says that Ancient Judaism (roughly by the beginning of the Christian era) was a religion of observance, which presupposes a law whose precepts should be obeyed for their own sake.

process of rapprochement between the two cultures in Brazil (Freyre 1986 [1933]:372-4).

The style is flowery, the language outdated, the words no longer politically correct, yet Freyre is drawing attention to a basic aspect of traditional Brazilian culture, with theoretical implications that bear upon the controversy surrounding the thesis of the Protestant ethic and the development of modern capitalism. The feast which, in a way, represents the rite in its pristine purity, having no other intrinsic meaning but its own realization, allows for the coexistence of several 'etiologies' or systems of explanations. This is the very process which lies at the root of Brazilian popular syncretic religions (and certainly of other forms of syncretism as well).[18] The civilization of the feast, as Weber himself was only too aware of it, finds itself in diametrical opposition to a civilization based on disenchantment and rationality. The pristine purity of the rite is equivalent to the pristine impurity of ideas. To a religion that centers on the feast, it is hard to apply a Weberian[19] concept of rationality as the quality of a thought that is conscious of itself in conceptual reflection and abstraction, which is the kind of thought of the economic rationality associated with the capitalist revolution.

Sacrifice and Trance

In the beginning there was the body. The body of the feast and of sacrifice, which, in the Candomblé and Xangô[20] is always accompanied by trance. As already stressed, in traditional Afro-Brazilian worship thought has a very markedly concrete character, being encapsulated so to speak, in images, in trance and the motions of dance. And in those religions blood is, as devotees say in their parlance, the supreme axé, (that is, that which causes and restores health and wholesomeness). No other element could express better than it the fullnes of life. Sacrifices, those transfusions of life, are primarily transfusions of blood.

---

[18] The idea of the ritual and the feast as susceptible to explanations by fairly different 'etiologies' is no doubt a central one in Gerhard von Rad's interpretation of the Genesis (von Rad, 1968).

[19] Or, for that matter, Hegelian.

[20] Candomblé is the name given, mainly in the cities of Salvador da Bahia and Rio de Janeiro (from where if has diffused elsewhere in Brazil) to the more traditional variety of Afro-Brazilian religion. Its equivalent in the area of Recife is called Xangô. Bastide 1978, 2000 are still considered as standard treatments of the subject. See also Motta, 1988b, 2001, 2003b, 2005. I use the two words interchangeably throughout this paper.

An example will now be furnished based on (abridged) field notes taken, some years ago, in the archetypal house of Badia, a senior Xangô priestess of the area of Recife:

> October Feast
> Smell of condiments in the kitchen. Boiling water. Talk about many sacrifices. Raminho describes his recent voyages. He tells people he will soon fly to Rondônia, in the far west of Brazil, to initiate there a few children in sainthood. First goats brought in. Frame song for sacrifice: Exu xoxô. Ejé balé cararô. ('To thee alone. May the blood shed upon the earth give us peace.'). One goat, four chickens. Tiriri lonan! Mighty on the road! Casualness. Petty talk goes on. Salt laid on blood. Casual talk inside and outside the sanctuary. Dancing steps. (The gracile motions of the young hunter). Atotô! We fear thy might! Excitement grows. Smell of boiling feathers. Pungent smell of rum. Blood flows and runs. Four birds. Vital forces on the loose. Unbearable smell of blood. Butchers (junior members of the priestly staff) in the backyard. Sun rays through glass tiles. Astronomical noon. Badia prays for health, happiness, everything. A pool of light red blood. 'Blessed be our Lord Jesus Christ!' Badia comes in.'Say, Raminho, what happened?' 'I tell you, we can't go on killing right now. The floor is much too oily, the sun is much too hot. We're tired. Let's have dinner and rest. We'll begin again later in the afternoon. Say, Roberto, would your doctor allow you to drink a glass of beer?'[21]

Heads of animals are severed in sacrifice with a sharp knife. Victims are bled to the last drop of the blood that falls upon either the stones, or the heads of people, or both. Animal heads are first superimposed upon the heads of the devotees. Then they are put on the altar. Trance is *de rigueur* on such occasions. The faithful also present them with the symbolic gift of their own bodies, for the gods to inhabit and make themselves manifest in, especially in the trance and dance states. Boundaries vanish between the extremes and the middle term of the equation of holiness. I can hardly find in my notes an instance of sacrifice without at least a sketch of that commingling of heads. Sometimes people, even non-initiates, kneel and rub their heads against those of goats and rams (if the latter are properly immobilized); at other times the presiding priest touches the human and animal forefronts; but even if all such gestures were omitted, which they seldom are, sacrifices would still be offered for, if not precisely on, the heads of people.

---

[21] Notes taken on October 17, 1981.

The following example comes from of a huge obrigação[22] held, on October 8, 1977, in the house of Dona Severina -- one of the many priestesses of that name in Recife. The slaughter comprised four four-footed animals (a young he-goat, two rams, a she-goat), plus, to put it mildly, many birds. Among other things, this is what is written in the researcher's notebook:

> Head of goat severed . . .deposed on altar . . . First ram; the whole congregation (ladies first) touch the head of the beast. A female devotee falls in trance. Ram bled to the last drop. Head entirely severed. Head of beast placed upon head of woman. Now head deposed on the stone. Devotee has blood in mouth and hands. Another ram for a blond man. Same treatment. . . . She-goat . . . Blood flows in gushes. Female devotee falls in trance. She sucks blood. Head of goat upon her head.

Sacrifices are good to eat, as the meat of the chair of the victims is quickly changed into food for the devotees and their friends. Yet, it also good to think and indeed it is simply good to be. It attempts to transfer the devotee from this world, the world of appearance, of instability, and death, to another world, the world of truth, of permanence, and everlasting life. The saints, as principles that transcend everyday experience, are the very foundation of reality and identity. On the other hand, they appear to have no other reality but that of ever renewed human movements, of running blood, of animals 'dropped' on stones, of singing voices and dancing bodies.

Trance is a variety of thought, but it means at the same time less and more than plain conceptual thought. It is never clear, logical and rational.[23] But it also transcends thought. There is in it such an overflow of representation, symbol, and identity, that the cognitive and affective faculties of the devotee become saturated with the strength of the message and no longer process it in a usual way. He is much too full of, and overwhelmed by, his saint and the enthusiasm of his group to be able to articulate his feelings in a rationally and conceptually clear way. Ecstatic rapture appears as a solution for the opposition between the apparently unlimited reach of symbol and communion and the narrowness of the devotee's ability to handle and assimilate it.

The dancing body has its own time, which is certainly not money, as demonstrated in the following instance:

---

[22] Sacrificial ceremony.

[23] And if it attempts to be, as it happens in Umbanda (see below), it stops being trance.

The dance, due to start at four o'clock p.m., actually began at eight. I waited, while the air became thick with the movements, the noises, the odors (that blend of sugar and sweat that seems to represent the hallmark of the Xangô) of adults and children who talked, ate, drank, watched television. It was like the threshold of a different world, so separate from everyday life that one only gains access to it through the sacrifice of time. Tonight I attended a kind of chamber concert. People were relatively few; they all knew, and were emotionally attuned to, one another; there was no competition for status or precedence. I decided to make no use of my camera. I was aware that Dona das Dores would be vain enough to enjoy being photographed at certain crucial moments, but the flashes would alter this perfect psychological environment. My interest has recently shifted to the mythological aspects of the cult. What do the songs mean, what message do they convey? Manoel tells me that no single person in the whole of Brazil can decipher their original meaning. 'But can't you tell me why this song mentions Ogum Oniré? -- I tell you: it celebrates the Ogum of the Onirés, a people from Africa. Don´t you hear the word corô? The Africans mispronounced the word coroa (crown in Portuguese). Listen, Roberto, Ogum is strong, strong, he wears a crown, he is a champion, he wears a crown.'

This represents a gain for Xangô. If the Yoruba of West Africa attributed a meaning to the words of their songs, that was due to the fact that only some professors of linguistics invent nonsense syllables. Here they don't invent them. They go beyond. They dispense with their literal meaning. Perhaps this loss is quite regrettable. But they have kept the deepest meaning of the songs. What does it help knowing that Ogum is the lord of the Yoruba town of Iré --Ogum Oniré, Ogum Alaré-- if I don't know first that he is the winner when he fights? They know who Ogum is. They --and at this moment I belong with them-- reconstitute the living meaning of the saint. I now understand the importance of the long expectation, of the small gestures before the dance. Everyone had time, during the early anti-climax, to impregnate himself with the feelings and emotions of the rest of the congregation, to slowly coalesce into a single group. When the music turns to Obaluaê, do I need a special mythological knowledge in order to understand the dialogue between the father and the choir of the dancers? Obaluaê—Lord of the World—cries, claims, proclaims his might. Atotô! We respect thee! We have quit time, yet with measure and restraint. Trance is not anarchic, nor is it dispersive. Rather we are subject to the dispersion, to the nonsense of everyday life. Having quit the word of appearances, we are now, to borrow the words of the ancient Greek philosopher Parmenides, on the Way of Truth. History has been momentarily overcome as now there is no time.[24]

---

[24] Notes taken on April 7, 1975.

Hopefully the preceding examples have helped to demonstrate the central thesis, or one of the central thesis of this paper. In the beginning, thought was expressed in the motions of dancing. Conceptual systems were implied in gestures. The motions of the body were not subject to time as a measure of productivity, but followed their own rhythm. This changes, however, with the growth of the market and with the process of acculturation, which applies the rationalism of Western culture to other cognitive systems. It would, of course, constitute a task going well beyond the scope and limits of this paper to give a complete account of how change and modernization have overtaken Brazil. Thus I will be here limited to three specific mechanisms.

From Baroque to Cyber

There is, first of all, the inherent dynamics of economic modernization. Thus about twenty years after the last ethnographic episode related in this paper, I wrote in my research notebook[25] that that very section of the city, where I used to do field work for my Ph.D. dissertation, as it was the greatest biological and cultural reserve of the Xangô religion of Recife, had changed beyond ethnographic recognition. Not without pain, I noted the advance of modernization during the last 20 years. There are broad (if poorly paved) new roads, indeed one of the new contour roads of the city crosses the district, entailing the intrusion of many nightclubs, dancings and cheap hotels. There are also many new blocks of public housing, the likes of which can be seen everywhere in the world, including the 'cités' of the Parisian *banlieue*, the council housings in England and many apartment houses in Eastern Europe. On a Sunday afternoon I can see a multitude of loudspeakers. In a very high tone, they broadcast kitsch Latin-American and international music. I am now many light-years away from my terreiros.[26] For there is no common measure between these people, whom I see dancing like people dance on Saturdays or Sundays in similar neighborhoods elsewhere, New York, London, Paris, Bucarest, who undergo modernization, indeed 'ultramodernization'[27], without,

---

[25] For my research is an unending quest and is in a way the task of my whole life.

[26] Terreiros are the shrines of Xangô and Candomblé.

[27] As the Latin prefix ultra has a general meaning of beyond, ultramodernizattion can be well understood as exacerbated modernization, which, in a fully metaphorical sense (indeed exacerbatedly so) might as well be called the cybermodernity. I follow Ivan Varga when he writes 'Postmodernity, in my view, is an excess of modernity and at the same time the expression of its crisis. It is not an absolute break with modernity, as modernity is not an absolute break with

however, succeeding in assimilating the core of its basic attitudes, and the congregations of the Xangô, of which I was such an intimate observer and participant barely 20 years ago,[28] very much in the same neighborhood, with their often grave and tragic music that sprang from what a people possesses of most intimate and deepest.[29]

Sheer 'infrastructural' mechanisms are perhaps the most powerful ones in leading the baroque dancing body to change into the cyber[30] dancing body. But two tendencies of a more intellectual kind also deserve mention in the recent transformation of the Afro-Brazilian cults and their underlying conception of the body. I refer, first, to the Umbanda movement, which appeared in the middle years of the 20th century, intending to 'purify' and to 'civilize' the African-influenced religions of Brazil by subjecting them to a process of reinterpretation based in large part on theological principles drawn from the writings of Frenchman Allan Kardec[31] (1804-1869), who, under the influence of Christian, Hindu and other mediumistic and reincarnationist beliefs became the systematizer or 'codifier'[32] of Spiritism.

The new movement was neither monolithic nor uniform. But some traits are common throughout the whole range of Umbandista centers, such as the classification of spirits in lines and phalanxes reminiscent of the medieval speculations about angels and demons. A belief in evolution, basic in the doctrine of Kardec and his Brazilian followers, has also been adopted by the movement, which welcome all spirits if they were willing to conform to the new rules of interpretation. Umbandistas have no objections to dealing with orixás (the saints of Candomblé which some of them would denominate pretos velhos, that is, 'old black people'), caboclos (Indian spirits), mestres (healers), ciganos (Gypsies) and sundry

---

premodern conditions; societies we call modern still contain elements of premodern thought that are often in contradiction with the dominant social structures and relations' (Varga, 2005:210).

[28] Let us not conclude, however, that old-fashioned **terreiros** vanished entirely from that nieghborhood. Although increasingly subject to changes and interpretations 'orthodox' congregations are still perfectly discernible, side by side with **Umbanda** centers and indeed with a proliferation of Pentecostal churches

[29] This paragraph follows notes I took on September 1st, 1995.

[30] Expressions such as **baroque** and **cyber** are, in this context, primarily metaphors, the former for the kind of traditional society that used to exist in colonial Brazil and which was, a matter of fact, largely influenced by the baroque Counter-Reformation culture of Latin Europe, and the latter for the excerbated modernity of the turn of the millenium, largely charaterized by a real cybernetic revolution that spared no region of the planet.

[31] Allan Kardec was actually the *nom de plume* of Hippolyte Léon Denizard Rivail.

[32] Often designated in Brazil simply as 'O Codificador'.

other 'entities', provided they accept the ethos of 'development' which constitutes a hallmark of Umbanda.[12] The primitive and backward are at the same time capable of progress and they will eventually catch up with the advanced.

Umbanda also redefines the ecstasy of Candomblé in mediumistic terms, changing it into a kind of verbal possession in which the ordinary personality of the medium is replaced by that of an 'entity' prone to offer advice and prescriptions useful for what ails people morally and/or materially. A further opposition between the Afro-Brazilian cults and Umbanda stems from the former's orientation toward the relief of the sorrows and afflictions of daily life, while among Umbandistas (who, however, are certainly not opposed to the more tangible and immediate aspects of religious consolation) one can discern the beginnings of a sense of history, entailing a project which presides over the evolution of the spirits and the 'development' of individual devotees.

Umbanda appears, therefore, as a movement which, in Weberian parlance, tended to rationalize the old Afro-Brazilian religion, through the adoption of a systematic (or 'codified') theology, through a new interpretation of trance[33] and other such features. If in the beginning there was the body, the New Testament introduced by **Umbanda** brings about the word. This is very much what (in spite of a somewhat naive trust in a social class paradigm of interpretation) Roger Bastide thinks of the new tendency:

> The spiritism of Umbanda corresponds to the rise of a racially heterogeneous low social class and to its tendency to ascend in the context of a competitive industrial society. Likewise it signals the transformation of intellectual thought under the influence of an environment marked by an all-powerful rationalization, for this kind of spiritism represents the passage from a system of symbols to a system of concepts (Bastide, 1975:195).

Little did Umbanda's founders[34] anticipate, however, that an increasing

---

[33] There is a vast literature on Umbanda, written by Brazilians and Brazilianists, in Portuguese, French, English, German and Italian. Beside the writings by Bastide mentioned in this paper, Ortiz, 1978, and Brown, 1984, are standard works concerning Umbanda. Motta, 1988a, can also be read with profit concerning the differences between Candomblé, Umbanda, and other Afro-Brazilian or Afro-Indo-Brazilian cults.

[34] Who were especially active from the 40's to the 70's of the 20th century.

number of devotees of the Afro-Brazilian groups, with an all-important technical assistance of some anthropologists and sociologists, both Brazilian or foreign, would, in the name of African authenticity, produce an alternative to compete with, and perhaps supplant, theirs in the Brazilian religious market.[35] To many of them the avowed syncretic integration of Umbanda has come to be viewed as a defilement of an idealized pristine purity of African beliefs and rites. To the extent that it was intended as a defense and 'spiritualization' of Africanhood, Umbanda, although it should by no means be considered as dead, is finding itself outdated and redundant.

The Holy Alliance

Scholars studying Brazilian religions often tend to evaluate religious movements according to their conformity to the criteria they consider as representative of modernity. Afro-Brazilian religions, in spite of their conspicuous sacrificial character, agree with a certain modernity by both their practical rejection of the notions of sin and guilt[36] and by being religions of the 'oppressed'.[37] Therefore, thanks to the writings of sociologists and anthropologists, the Candomblé religion was invested with highly rationalized theological reinterpretations. Congresses and conferences, attended by both researchers and religious leaders, function as ecumenical councils during which faith is defined and proclaimed.

Brazilian social scientists have not limited themselves to a value-neutral and detached study of religious institutions and changes. In fact, they have often taken on a managerial role in the matter, on the grounds that it is incumbent upon them, as anthropologists and sociologists, to

---

[35] The very name Umbanda, which in large and fashionable circles of Afro-Brazilian devotees in the early 1970's still represented a dignified way to refer to Candomblé and Xangô, has come some 30 years later to acquire a negative and politically incorrect connotation as a kind of Uncle Tomism.
[36] For all practical purposes, the notions of sin and guilt, such as they exist in Roman Catholicism, are replaced by the awareness of a sacrifice or a gift owed to the saints, or to a specific saint. This is indeed a basic apprehension of the Afro-Brazilian religions, expressed by the Portuguese word obrigação (duty, obligation, debt). Every cultic act is conceived as the accomplishment of an obrigação. Cultists view with great naturalness the manifestations of the sex drive (in its whole gamut) and are far from enjoining the practice of chastity or abstinence , except on some precise ritual circumstances.
[37] Although associated with the descendants of African slaves, Candomblé and Xangô heartily welcome the adhesion of Brazilians and foreigners of other origins.

define modernity and to discern its appropriate features. They act as if social science had the task of providing for the interim management of the religious sphere.[38] A new syncretism results from the way social scientists have taken possession the Afro-Brazilian religions and invented for them an entire, highly rationalized theology.

A holy and scholarly alliance has therefore been established in Brazil between Afro-Brazilian religion and the social scientists that define and represent certain of modernity's values. Scholars have, moreover, often been well aware of the role they were playing in this creation --for that is what it is-- of an autonomous Afro-Brazilian religion. Roger Bastide, for instance, expresses himself, in a little text published 55 years ago (i.e. even before his major works on African religions in Brazil), as follows:

> I recall that Anísio Teixeira, with his characteristic lucidity and intelligence, reproached us—Ramos, Herskovits, Pierson, Carneiro and myself—with having strengthened the Candomblé and thereby hindered the assimilation of the Black in the north-east of Brazil to Western culture. He was quite right. The pais-de-santo[39] use our work in order to comprehend Africa... (Bastide, 1953: 521).

Thus researchers have given Candomblé, which has, of course, always been, in itself, a religion with a well-structured theological system, its own rites and myths and its priesthood, which enable it to become a self-sufficient religion, a religion for itself, with an enlightened awareness of its unique and independent status among other religions which compete with it for converts.

The Xangô, the Favelas and Modernity

According to research done in the late 20th century, members of the Afro-Brazilian religious groups, in the area of Recife, belonged, as a general rule, to the informal sector of the economy.[40] And, giving that word the general meaning it has in Brazilian Portuguese, that is, connoting popular housing, inhabited by persons of low income and not fully integrated into the dominant sectors of the economy, who thus resort to unconventional and informal activities, it can be said that the adepts of Candomblé, Xangô, and other varieties of the Afro-Brazilian religions

---

[38] Until, presumably, the final disappearance of religion.
[39] Pais-de-santo are the priests of Candomblé and Xangô.
[40] On this topic, see Motta & Scott, 1983; Motta, 1984.

(and, for that matter, a considerable part of the converts to the Pentecostal churches) live in favelas.

Around 1980[41], the unemployed plus the underemployed[42], comprised almost 30% of the total labor force of Recife. The economic environment used to (and to a large extent still does) push people out of, and did not allow them into, the full circuits of a monetary, abstract economy. Therefore concreteness was at a premium, that is, whatever represented a concrete opportunity for obtaining gained or transferred income was functional in such setting. This pervaded every institution of the area, including the religion, which suffered an environmental pressure toward keeping or adopting more than purely abstract symbols in its ritual and organizational structures.

Formal, Western style education represented a clear advantage in this setting. But while it was not unavailable to those who could afford it, it was an expensive commodity. Those who were unskilled according to the requirements of the dominant economy had to find subsidiary or marginal niches. Suppose that we are dealing with services of a religious kind. In the Roman Catholic or in the historical Protestant denominations such as the Presbyterian Church, those services will require, in order to be effectively and legitimately offered in the 'market', educational curricula that represent variants of the Western patterns of formal education. Here precisely lies the advantage of the Xangô and related cults and, at that, of the Pentecostal churches.[43] They do require education in the sense that they require strenuous training under specialized tutors. But while this process involves costs[44]—nothing in Xangô is done without some transfer

---

[41] This time corresponds roughly to the period during which the author of this paper did research for his PhD dissertation (Motta, 1988). For a more recent picture see Motta, 2002.

[42] That is, people with an annual income equal to, or below, US $308.00 at 1980 values.

[43] The Afro-Brazilian cults and Pentecostalism, albeit in different ways are, as a distinguished Brazilian anthropologist put it, 'religions of participation' (Ribeiro, 1982), both on account of the direct contact with the supernatural (mainly through trance) and of the absence of any too pronounced separation between 'clergy' and 'laity', the members of the latter being able to accede to the former with at least relative ease. This aspect was also stressed by Motta (2003a). The same subject was treated by Frase (1975) --who does not hesitate to qualify the Pentecostal as 'acculturated Protestants'-- in a remarkable study that has so far been available only as a PhD dissertation.

[44] 'No Candomblé não se faz nada de graça', nothing in Candomblé is done for free, as Melville Herskovits used to say, quoting his informants. (Herskovits, 1966:253).

of income—its price is much more accessible than that of the qualifications acquired through formal, Western education.

Western style education does not act only as the transmitter of certain skills. It also acts as the conveyor of patterns of abstract thought and reasoning, which, in their turn, are largely functional to the infrastructural requirements of a modernized, industrial economy. Conversely, the absence or the weakness of Western patterns of education mean, in our setting, that modes of thinking that above all stress the use of matrices embedded in concrete symbols or images have persisted and have not quite given way to an abstract, science- or rationality-oriented frame of mind.

Candomblé and Xangô, as previously pointed out in this paper, are very concrete religions[45] in that their theological matrices are conceived by the devotees as embedded—one might say as inhabiting—concrete, visible, touchable things and states. They are, to put it in Victor Turner's (1975) vocabulary, markedly 'iconophilic'. The visible, touchable, hearable enactment of its theology practically excludes an abstract dogmatic system --by no means are they 'religions of the book'.[46] Whatever the religious and artistic values and traditions of which the cultists have been the bearers, they preserved the concreteness of forms of thought unmediated by 'rationality' and 'calculus' (in the Weberian sense) because they had practically no alternative, given their exclusion from the dominant sectors of socio-economic activity. Thus the Afro-Brazilian religiousness can be considered as an inverted mirror image of the abstractive ways of thinking brought about by the people-dispensing economic development in the area.

The religion is thus coherent with its social and economic setting. Nevertheless, a caveat must be strongly sounded in this context. There is no possible way to deduce the presence of this religion in this setting from any kind of infrastructural requirements. Candomblé and Xangô are contingent phenomena. They result from no universal laws, but from a specific and concrete history. Moreover, they go well beyond all kinds of economic or, at that, political conditions in that they are also the expression of a deep sense of ethnic and personal identity. The Afro-Brazilian religion is thus at the same time affected by, and resistant to, modernization. Its demise has been announced many times by social scientists and other analysts. The future is not ours to see. But it is likely that some nuclei, composed of particularly homogeneous people from a

---

[45] Or rather as two branches of the same religion.

[46] Or, at least, they were not 'religions of the book' until affected by the theology of Umbanda and by the learned treatises of anthropologists and sociologists.

social and ethnic standpoint, will continue, through the coming generations, to practice their religion and to express their identity reasserted in feasts that take place along the year, in sacrifices, dances and trances, in which the individual devotee is fused with his group and his god.

Yet, modernity has never stopped corroding the native patterns of thought of the Afro-Brazilians,[47] as we have seen by the examples of the Umbanda movement and of the rationalized theology social scientists have been trying to give the cultists. Indeed, living in the same country, the same society, speaking, in their daily lives, the same language of other Brazilians, devotees could hardly escape the influence of other systems of thought whether in the guise of intellectual speculations or embedded in actual economic behavior. The process of erosion and change gains momentum, around the turn of the millennium, with the impact of modern electronic technology (especially cyber-technology) on the everyday life of favelados.[48] This has been an indeed quick development that revolutionized life in the favelas and therefore could not fail to affect the Afro-Brazilians. Mobile phones became quickly commonplace.[49] CDs also quickly became normal, allowing, among other things, enterprising *pais-de-santo* to record and market the canticles of their religion, which, however, are more and more subject to the requirements of the brisk time of profit, so different from the time of the body associated with sacrifice and trance. Nevertheless it is not probable that even those new entrepreneurs have assimilated to the core the values and attitudes of a modernity associated with rationality. This is the essence of the Brazilian drama.[50] That is, the persistence, betwixt and between, of a large informal sector, characterized by the simultaneous presence of survivals of the baroque body and the paraphernalia of the cyber body.

Concluding Remarks

The Afro-Brazilian religions have increasingly tended to be caught in the iron cage of rationality. This process was inevitable in a country in

---

[47] Which, from the beginning, represented a blend of, among other ingredients, Yoruba beliefs and baroque Roman Catholicism.

[48] Inhabitants of favelas.

[49] The author of this paper remembers he saw a mobile phone for the first time in January of 1993.

[50] Not wholly unlike the **Asian drama** studied by Myrdal nearly 40 years ago. Optimists may hope that the drama will end, with but little upheaval, with the advance of Western style education and a significant drop in the birth rate.

which, despite many ethnic, social, and economic peculiarities, it is, on both the intellectual and technological level, the modernized sector that prevails and expands constantly. Despite the survival of tradition among some restricted nuclei, the very conception of the body changes in those religions. Thought is no more encapsulated in dancing bodies and radiant images. To the contrary, the body and its movements change more and more into signs or icons (not unlike computer icons) of beliefs duly systematized according to Western logic.[51] Having in view the Brazilian experience, the conclusion of this paper can perhaps only in degree be different from those of Max Weber at the end of *The Protestant Ethic and the Spirit of Capitalism*. He refers to

> the tremendous cosmos of the modern economic order. This order is now bound to the technical and economic conditions of machine production which today determine the lives of individual who are born into this mechanism, not only those directly concerned with economic acquisition, with irresistible force. [...] Fate decreed that [... it] should become an iron cage (Weber, 1958: 181).

This is above all sheer ethnographic observation, seizing, right at 'the summit of history' the basic tendency of our time or of a past which has not yet past. The author of this paper disclaims any involvement with political projects aiming at either maintaining or reversing this situation. For, committed as he has been, in this and other papers, to a detailed account of the Afro-Brazilian religious sentiment and expression, he has tried to study the facts with the utmost objectivity. The strictly scientific aim of his work shows that he is in no way associated with the controversies between the metaphysicians of the matter and those of the spirit or, for that matter, between those of modernity (or ultramodernity) and those of the utopian dream of a return to a golden age prior to the realm of commodification and rationalization.

---

[51] And the project of a return to the Baroque and imagistic dancing body is all the more unfeasible as this kind of project is itself impregnated with rationality and abstraction.

References

Abramowski, Günter. 1966 *Die Geschichtsbild Max Webers: Universalgeschichte am Leitfaden des okzidentalen Rationalisierungsprozesses*, Stuttgart, Ernst Klett Verlag.
Bastide, Roger. 1953 '*Carta Aberta a Guerreiro Ramos*', in *Anhembi* (São Paulo), vol. XIII, no36.
—. 1975 '*La Rencontre des Dieux Africains et des Esprits Indiens*', in Roger Bastide, *Le Sacré Sauvage*. Paris: Payot. Pp. 186-200.
—. 1978 *The African Religions of Brazil*. Baltimore, Johns Hopkins University Press.
—. 2000 (or. 1958) *Le Candomblé de Bahia (Rite Nagô)*, Paris, Plon.
Berger, Peter. 1986 *The Capitalist Revolution*. New York, Basic Books.
Brown, Diana De G. 1994 *Umbanda: Religion and Politics in Brazil*, New York, Columbia University Press.
Bultmann, Rudolf. 1984 new edition *Theologie des Neuen Testaments*. Tübingen, Mohr Siebeck.
Duerr, Hans Peter. 1998 *Nudité & Pudeur: Le Mythe du Processus de Civilisation*, Paris, Éditions de la Maison des Sciences de l'Homme.
Elias, Norbert. 1978a *The Civilizing Process*, volume I: *The History of Manners*, New   York, Pantheon Books.
—. 1978b *The Civilizing Process*, volume II: *Power and Civility*, New York, Pantheon Books.
Foster, George. 1967 *Tzintzunzan: Mexican Peasants in a Changing World*. Boston, Little Brown.
Foucault, Michel. 1984 *Histoire de la Sexualité*, Paris, Gallimard, 1984.
Frase, Ronald Glen. 1975 *A Sociological Analysis of the Development of Brazilian Protestantism: A Study in Social Change*, PhD diss. Princeton Theological Seminary.
Freyre, Gilberto. 1986 (or. *1933*) *The Masters and the Slaves: A Study in the Development of Brazilian Civilization*. Berkeley: University of California Press.
Herskovits, Melville. 1966 (or. 1958) 'Some Economic Aspects of the Afrobahian Candomblé', *in* Melville Herskovits, *The New World Negro*, Bloomington, University of Indiana Press, pp. 248-265.
Hyppolite, Jean. 1983 *Introduction à la Philosophie de l'Histoire de Hegel*, Paris, Seuil, Jaspers, Karl
—. 1988 'Max Weber. Eine Gedenkrede' (or. 1920), in *Max Weber* (Gesammelte Schriften), München, Serie Piper, pp. 32-48.
Lyotard, Jean-François. *1979 La Condition Postmoderne*, Paris, Les Éditions de Minuit.

Marx, Karl. 1967 *Capital*, vol 1. *A Critical Analysis of Capitalistic Production*, edited by Frederick Engels, New York, International Publishers.

Motta, Roberto. 1984 'Xangô e Estratégias de Sobrevivência', *in* Inaiá Carvalho e Teresa Haguette, orgs., *Trabalho e Condições de Vida no Nordeste Brasileiro*, São Paulo, Hucitec.

—. 1988a 'Indian-Afro-European Syncretic Cults in Brazil: Their Economic and Social Roots', *Cahiers du Brésil Contemporain* (Paris, Maison des Sciences de l'Homme), no. 5, pp. 27-48.

—. 1988b *The Xangô Religion of Recife, Brazil*. Ph.D. dissertation, Department of Anthropology, Columbia University in the City of New York.

—. 2001 'Ethnicity, Purity, the Market and Syncretism in Afro-Brazilian Cults', in Sidney M. Greenfield and André Droogers, eds., Reinventing Religions: Syncretism and Transformation in Africa and the Americas, Lanham (Maryland), Rowman and Littlefield, 2001, pp. 71-85.

—. 2002 'L'Expansion et la Réinvention des Religions Afro-Brésiliennes: Réenchantement et Décomposition', *in Archives de Sciences Sociales des Religions*, 117 (janvier-mars), pp. 113-125.

—. 2003a 'Religiões Éticas e Religiões Sacrificiais: Seu Crescimento Simultâneo no Brasil Atual', *in* Maria do Carmo Brandão & Antônio Motta, orgs., *Anais do VII Encontro de Antropólogos do Norte-Nordeste*, Recife, Programa de Pós-Graduação em Antropologia, CD, 12 pp..

—. 2003b 'Le Sacrifice Xangô à Recife', Social Compass, vol 50, no 2, pp. 229-246.

—. 2005 'Body Trance and Word Trance in Brazilian Religion', in *Current Sociology*, vol 3, no 2 monograph 1, March 2005. *Bodily Order, Mind, Emotion and Social Memory*, edited by Bianca Maria Pirani and Ivan Varga, pp. 293-308.

Motta, Roberto, & R. Parry Scott. 1983 *Sobrevivência e Fontes de Renda*, Recife (Brazil), Massangana.

Myrdal, Gunnar. 1963 *Asian Drama: An Inquiry into the Poverty of Nations*, New York, Pantheon.

Ortiz, Renato. 1978 *A Morte Branca do Feiticeiro Negro*. Petrópolis (Brazil), Vozes.

Rad, Gerhard von. 1968 *La Genèse* (French trans.). Geneva: Labor et Fides.

Ribeiro, René. 1982 *Antropologia da Religião e Outros Estudos*, Recife (Brazil), Massangana.

Sombart, Nicolaus. 1955 'Einige Entscheidende Theoretiker, Ursprünge, Henri de Saint Simon und Auguste Comte', in Alfred Weber (Herausg.) *Einführung in die Soziologie*, München, R. Piper & CO Verlag, pp. 81-102.

Sombart, Werner. 1913 *Der Bourgeois: Zur Geistesgeschichte des modernen Wirtschaftsmenschen*, Leipzig, Duncker und Humblot.

Tapié, Victor-Lucien. 2002 *Le Baroque*, Paris, PUF.

Tönnies, Ferdinand. 1978 *Community & Society*, Charles P. Loomis, transl., New Brunswick, N.J., Transaction Publishers, 1988, 298 pp.

Turner, Victor. 1975 'Symbolic Studies', *in Annual Review of Anthropology*, vol. 4, pp. 145-162.

Varga, Ivan. 2005 'The Body –The New Sacred: The Body in Hypermodernity', in *Current Sociology* vol. 53, no. 2, monograph 1, *Bodily Order: Mindd, Emotion and Social Memory*, pp. 209-236.

Weber, Max. 1958 *The Protestant Ethic and the Spirit of Capitalism*, translated by Talcott Parsons, New York, Scribner's.

—. 1978 *Economy and Society* (Guenther Roth & Claus Wittich, eds.), 2 vols, Berkeley: University of California Press.

THE PAST, PRESENT:
LAMALAMA INTERACTIONS
WITH MEMORY AND TECHNOLOGY

DIANE HAFNER

Abstract: In popular discourse in Australia, Indigenous authenticity is still at times regarded as vested in remote-dwelling people with a total commitment to seemingly timeless and unchanged traditional practices. That such people can be enthusiastic in their acceptance of technological innovation is frequently misunderstood as a sign they have abandoned their culture. These views fail to account for history, region, and transformations of social space, and the way in which communication and other technologies re-configure the space between the individual and their social world. Concentrating on the experience of one remote Indigenous group, the Lamalama of Cape York Peninsula, I discuss preliminary findings from museum-based research that frames technology and memory as collective forms of 'remembering'. To date, it appears for the Lamalama the past is labile rather than historically isolated. It is the source of possibility and interpretation as well as the location of mutually known 'facts', and exists in the present as well as being temporally distant. This mobility of mind and practice works through certain forms of collective memory, and is consistent with the view that modernist technologies are simply a part of the assembly of cultural tools by which the Lamalama produce and maintain a coherent identity.

In Australia, the popular perception of Indigenous people remains ethnocentrically focused on their supposed exoticism, and in various ways in the popular imagination, they continue to be perceived as deeply traditional and disinclined towards innovation (Attwood 1992, pp. iii-iv). Yet such groups have long demonstrated a pragmatic interest in the new as a complement to tradition. Here I report on approaches to electronic technology taken by one north Australian Aboriginal group, the Lamalama of Cape York Peninsula. I also discuss preliminary findings from a research project in which their access to ethnographic materials in museum collections act as *aide mémoires*.

I discuss notions of memory and the past, and briefly describe their importance to the Lamalama. I provide an overview of historical events that have impinged on them, in particular those causing them to suffer cultural discontinuities beyond their control. Memory, facilitated by access to museum materials, is being used by the Lamalama to create a contemporary identity that is consistent with the past and their obligations towards it. The research of the anthropologist Donald F. Thomson, who worked with their forebears in the late 1920s, is central to this. The present situation of the Lamalama including their interest in and use of contemporary technology provides the context for discussion. Drawing on the cultural context in which Lamalama acts of remembering occur, I suggest there is no disjunction between tradition and innovation in their uses of technology.

Research with the Lamalama

The research project I discuss brings the Lamalama into close contact with material evidence of their past stored in Museum Victoria, a public museum complex approximately 4000 kilometres distant from their home, in the southern Australian city of Melbourne. This evidence resides in artefacts Thomson collected in Cape York Peninsula between 1929 and 1932, and is most significantly a large collection of black and white images and his associated field notes, and some separate items of material culture. Methodologically, the project involves an iterative process of videoing groups of Lamalama people on visits to the Thomson and other Collections and during fieldwork in Cape York Peninsula. The information they generate is returned to the larger community after the museum visits.

In this process the Lamalama are participating firstly in a project that investigates objects of material culture as objects of relevance to themselves and secondly, re-invigorating an identity contained in what might otherwise be collections of 'lifeless objects' (Bolton 2003). There is not much direct Lamalama memory of the period, or of Thomson's visits. The current senior generation of Lamalama people were either infants or not yet born during the 1920s. The items contained in the Thomson Collection and the information about them are a source of evidence about past practice, although removed in time and space from their original contexts, and therefore disconnected from the meanings and purposes they held for their makers and users (Bolton 2003, pp. 43-44). Nonetheless, current Lamalama attempts to strengthen cultural identity depend on its contents. Museums, and by extension, the research and broader Australian community benefit from the research activities described here, through the greater precision in collection management that is facilitated by the information that Indigenous people such as the Lamalama provide. Such interests need to be declared from the beginning, but they are not the only outcomes of the project. The Lamalama have participated in research into the Thomson Collection for several years, and the granting of competitive national research funds means we are able to comply with their requests to continue and expand the work[1].

I am particularly concerned here to discuss the idea that memory in this context might be regarded as deriving from social actions in the present rather than pertaining only to direct, first-person recollection of past events. I recognise that constructions of the past resulting from Lamalama group interactions inevitably involve a process of negotiation between people and objects: firstly, between the members of the group as individuals, but also the objects as potentially contested sites of identity concerned with competing notions about the group's past. I do not focus on the latter form of negotiation here, although it may be a significant focus of the research as it continues. Rather, I concentrate on Lamalama interest in and uptake of 'modern' technology. The Thomson Collection contains objects with meanings potentially subject to manipulation by the Lamalama through a variety of such technologies. In seeking those meanings, they draw on what Zamora (cited in Wertsch 2002) identifies as the 'usable past', to produce a more integrated and economically viable identity than is currently available to them.

The Lamalama Past

The Lamalama are a small group of Aboriginal people or *pama*[2] who live on or close to their traditional lands[3] at Port Stewart on Princess Charlotte Bay in Cape York Peninsula (Figure 1). Over the last fifty years, they and their forebears have endured enforced removal by the state, ensuing separation of kin from their families, and ultimately reinstatement on their traditional estates or country (Sutton 1995) as a result of successful claims and negotiations under a variety of legislation, including Australia's *Native Title Act* 1993 (Cth).

---

[1] Australian Research Council Linkage Project LP06607418. I am indebted to my colleagues for their generous support in the preparation of this paper, which draws on a pilot study for the larger research project and joint discussions relating to the continuing research I describe.
[2] This Indigenous language term of self-reference used to identify Aboriginal people of the south-east region of Cape York Peninsula.
[3] The traditional country of the Lamalama runs south around Princess Charlotte Bay from Massey Creek in the north down to the northern boundary of Lakefield National Park (see Fig. 1).

Both the materials in the Thomson Collection and the technologies used by the Lamalama to interpret and elaborate on them for the purposes of identity production involve what Wertsch (2002, p.17) and others (see Bartlett 1932) describe as a process of 'collective remembering', rather than collective memory. Wertsch (2002) views memory as often called upon to produce an account of the past that can be harnessed for some purpose in the present. The act of remembering, and the people engaged in so doing, are not divorced from the contexts in which these occur. Remembering is therefore inherently social and cultural (Wertsch 2002). Creating a 'useable past' through memory usually serves certain needs in the present (political or identity needs in Wertsch's view), and such is the case for the Lamalama.

The Thomson materials provide a link to a past that cannot be accessed simply through the individual memory of even the most senior of group members, in part because of their age at the time of Thomson's visit, but also because of the historical impacts of settlement. At the time, Aboriginal people were subject to the terms of Queensland's *Aboriginals Protection and Prevention of the Sale of Opium Act 1897*. The immediate intention of this Act was to ameliorate the mistreatment of Indigenous people on the frontier of settlement in Queensland, although its impacts rapidly became quite different. This notorious Act effectively made Aboriginal people wards of the state and vested almost total control of their lives in the hands of the Protectors, usually local police officers or mission superintendents, until well into the twentieth century.

Thomson first visited the Port Stewart region in 1928, having gained permission from the relevant Protectors of Aboriginals in the Queensland capital city of Brisbane, and Cooktown, a regional administrative centre (Rigsby 2005, p.132). When he arrived at Port Stewart in 1928, he found a group of people he first described as the *Yinjinga Tribe* camped there (Rigsby 2005, p.132), some of whom worked on fishing vessels and nearby pastoral properties. In this they were anomalous, as few groups of Aboriginal people were free to live outside missions or reserves and direct their lives for themselves. During his time there, Thomson found this Port Stewart local group spoke at least four different Indigenous languages, although it was not then commonly thought such local groups were multilingual. What Thomson was observing at Port Stewart was the relatively early results of the dispossession of *pama* on the frontiers of settlement.

As the pastoral industry advanced, it took up lands for grazing purposes and forced the people of the lower Bay region to move north. By the middle of the twentieth century they and their descendants had consolidated as a kin group centred around Port Stewart, and by the late 1970s,were identifying themselves as 'the Lamalama people' (Hafner 1999). The group of people Thomson worked with were therefore mostly the forebears of the present Lamalama group, although they would have described themselves as members of separate language-owning clans rather than a single 'tribe' at the time he knew them.

Their seemingly autonomous life at the time of Thomson's visit did not go completely unnoticed. By 1961, after some years of lobbying the Queensland government by local pastoralists, the few dozen Lamalama people still living independently in the bush were removed. Police came to their camp and told them they were to be taken away for medical examination. They were escorted onto the government supply ship, the *Melbidir*, and removed to Cowal Creek, now Bamaga, some 300 kilometres away. From the boat, they looked back and saw their houses on fire and their dogs being killed. Under the terms of the 1897 *Protection Act*, they were not allowed to leave the mission at Cowal Creek, and most of the older people died there.

It was not until the 1980s, with the relaxing of the administration of Aboriginal affairs in Australia that numbers of Lamalama people remaining at Bamaga began returning to Coen, the town nearest to Port Stewart. There they took up residence among their few kin who had not been removed. By the end of the 1980s they had started to return to their traditional homelands at Yintjingga[4] or Port Stewart. They established a community that has continued to grow as a result of successful grants of land and other negotiations made under relevant legislation in the period since.

The Lamalama Present

The Lamalama currently maintain three separate communities on their Yintjingga country. A shifting population that responds to seasonal work opportunities and other demands and interests makes it difficult to be precise about the numbers, but the regional Lamalama population is generally between 100-150 people. Lamalama people also live in other nearby centres such as Cooktown and Hopevale. With people under 45 years of age making up the largest segment of this self-identified Lamalama Tribe, the need for young children to remain in school means

---

[4] Thomson's 'Yinjinga' is now generally rendered by the Lamalama and others as 'Yintjingga'.

that most people tend to reside in the nearby town of Coen during the week and travel down to Port Stewart on weekends, weather permitting, in this monsoonal region.

Yet few people can afford to maintain cars. Many roads in the region are unsealed, meaning expensive four-wheel drive vehicles are necessary, particularly for the cross-country travel required to access the remoter parts of Lamalama country. Such trips are consequently infrequent. Overall, Cape York Peninsula is an economically depressed region, with little commercial enterprise apart from the declining pastoral industry and isolated pockets of low-scale tourism. There are few jobs available, and even fewer of these are available to Aboriginal people. Some men work seasonally on local cattle properties, but such jobs are increasingly scarce.

In this context, alcohol consumption and attendant social disruption are continuing problems to them and a source of their interest in the possibilities offered by the Thomson Collection research. Senior Lamalama people view the Collection as a means by which to strengthen the cultural traditions interrupted or diminished by their contact with white settlement.

The Past in the Present

This demonstrated interest in how the technologies of modernity can be used to mediate 'traditional' identities and current possibilities is nothing new. In popular perceptions, the 'real' Aborigines are those remote-dwelling people who live by subsistence strategies unchanged by time or the impacts of European settlement (Attwood 1992). Yet there is a long history of Indigenous groups in Australia and elsewhere appropriating the technologies of modernity for their own purposes. For example, Sharp (1952) described the uptake of steel tools over stone implements among the Yir Yiront of Cape York Peninsula, in a process repeated endlessly on the frontiers of settlement. More recently, Morris (1989) has described Dhan-Gadi use of European domestic materials as a form of bricolage, while on Australian television in 2001, the program *Bush Mechanics* (Australian Broadcasting Commission 2001), followed a group of young Walpiri men from Central Australia as they applied alternative functionalities to standard mechanical items to achieve their desired goals.

To view Aboriginal authenticity as residing in a set of unchanging cultural practices is to set them apart from the rest of human experience. As Wertsch (2002:11) points out, 'to be human is to use the cultural tools, or mediational means, that are provided by a particular sociocultural setting', and the Lamalama are among those Indigenous people who are enthusiastic about using electronic technology to mediate between their past and present identifications as a group.

The sociocultural setting in which the Lamalama now live includes formal, state-based education, employment in the mainstream economy, kin relationships, and cultural practices such as hunting alongside the use of automobiles, telephones, and computers. Lamalama identity is currently mediated through all these means, by which the past is drawn into the present, and mobile phone use is the most obvious evidence of this claim. A mere nine people out of the current population of approximately 60 adults in the Coen-Port Stewart group do not own mobile phones. Of this group of 60, 45 are in the 25-45 age group, all of whom own and use mobile phones. Children form approximately 50 percent of the overall population of this core group, and all Lamalama children in Coen currently own and use mobile phones (S. Bassini 2006 pers. comm.)[5].

A number of the younger adults in the 25-45 age group are familiar with computers and use them regularly in work situations, but only one young Lamalama woman owns a personal computer that she uses at home[6]. Adults in the over 45-age group are generally not completely literate or interested in acquiring personal skills in computer use. Children learn to use computers at primary school in Coen, and some of the Lamalama are participating in the Computer Cultures[7] project run by Cape York Partnerships, an Indigenous community development organisation that supports *pama* endeavours to move beyond passive welfare recipience and toward social recovery (Cape York Partnerships 2005).

Indigenous tutors in the program engage students in activities such as educational games and projects to strengthen cultural identity and teach skills in use of the internet and electronic multi-media tools. Family biography activities are a popular means by which tutors lead students in the use of digital cameras and video to film their elders talking in Indigenous languages about important social and cultural matters. Tutors and students alike learn to produce materials that can be disseminated

---

[5] Ms Seppi Bassini, Coen, 27 January 2006.

[6] Most of the Lamalama cannot afford to buy personal computers.

[7] A primary goal of the Computer Cultures project is building higher Indigenous expectations for teaching and learning, with the aim of reforming the delivery of education for *pama*. It focuses on the transmission of knowledge through activities that engage families in supporting their children's learning. Recent activities include the Coen Family Archive and Coen Family WebPages, 'two systems designed by the project to support cultural transmission' using 'modern technology to assist families to share their knowledge and culture with the younger generations' (Cape York Partnerships 2005a).

through the internet or translated into both text-based and electronic formats such as PowerPoint presentations (S. Bassini pers. comm. 2006).

Most of the Lamalama people are thus familiar with the potential of computers and other technology, and what computers offer to them and their future is recognised and differentially valued. This familiarity with the possibilities of digital technology is notable given the degree of access to telecommunications available to the Peninsula community in general. Unlike metropolitan centres, where citizens have access to Global System for Mobile Communication (GSM) networks, people on the Peninsula are restricted to satellite phones or Code Division Multiple Access (CDMA) technology, and not all mobile networks in Australia operate in this format. Indeed, most parts of the Peninsula are covered only by the CDMA network operated by Telstra, the largest and partially privatised telecommunications network that is the country's main telecommunications infrastructure provider (Commonwealth of Australia 2002, p.36). It is worth noting that most of those who actively use computers are women, which probably reflects the nature of work available in the region. It is generally young women working in the few offices in Coen who have access to training and use of computers, whereas young men are more likely to pursue 'traditional' employment as skilled labourers in the cattle industry, although this is not exclusively the case. Such jobs are regarded as 'traditional' pursuits because these men are following in the footsteps of their fathers.

Mixed experience of this kind fits with 'postclassical' experience, a term coined by Sutton (1998) to describe the way in which some practices associated with the long stable cultural period before settlement have been maintained, while others changed as a result of contact with invading cultures. Usually applied to changes in social organisation, the notion of postclassical cultural practice is useful in this instance to describe the way in which people such as the Lamalama utilise modernist technologies to access their past and maintain it as part of their current identity. Despite their familiarity with audiovisual and telecommunications technologies, practices thought of as definitively 'cultural' or 'traditional' remain an important part of everyday Lamalama routine. Hunting, fishing with handmade spears and collecting plant materials for customary purposes are among such activities and represent the material dimensions of their postclassical identity as Lamalama.

Traditional Beliefs and Contemporary Identity

This identity is grounded in a belief system that is maintained in the face of contact with all the technologies of modernism, ranging from tools and implements to the dissemination of abstract ideas and knowledge through broadcast and electronic media. Yet the Lamalama continue to maintain there is an appropriate way to behave that defines relationships to both people and the environment. They talk about this as Pama Law, or 'Lamalama way', and it forms an integrated system of rules that locates the individual within a structured system of relations with kin and country, and a hierarchical relationship to two categories of numinous beings: the Story, and the Old People. The Story are the sacred creator beings whose actions made the world and its environmental features. Old People are deceased relatives who, though incorporeal, remain after death as animate social actors in the lives of their relatives.

The major Story for the Lamalama, *Marpa Haminhu* the Wind Story is located at Cliff Islands, a set of small islands off-shore from Port Stewart. Like the lesser Story located at sites on the mainland, *Marpa Haminhu* has the power to affect human lives in adverse ways, unless the appropriate rules for behaviour are followed. The Old People too have the power to impose themselves on the lives of the Lamalama. In Lamalama cosmology, individuals return to their clan country at death, where they are charged with the responsibility of regulating human actions within the landscape. Inappropriate action can be punished through the mechanisms of *kintya* or law in relation to the sacred sphere. Both the Old People and the Story are capable of punishing poor behaviour on country. Their actions do not apply exclusively to the Lamalama and may be extended to anyone behaving inappropriately on Lamalama country.

The Old People are also capable of more extreme forms of punishment, as evidenced in the condition of *puuya*, an illness that is associated only with punishment by the Old People for wrong-doing against their kin or 'countrymen'. To bring about *puuya*, the Old Person responsible must reach into the body of the victim and touch the heart or other parts of the alimentary tract, which brings on stomach pains and heart tremors. Lamalama fear this condition for the terrifying experience itself, but also because there is only one cure and failure to apply it will result in death. It is in this context of a cultural identity that accommodates a continuing tradition of specific belief, framed within the modernities of mechanical and electronic technologies that are available and readily applied by them, that the Lamalama approach research into the Thomson Collection.

Thomson photographed a variety of activities at Port Stewart, including burial ceremonies (Rigsby 2005, p.133) and more routine daily tasks. He used Thornton Pickard glass plate cameras despite the fact they were cumbersome to transport (Allen 2005, p.46-7), taking around 2,500 images in Cape York Peninsula, including approximately 500 at Port Stewart. His glass plate negatives still produce images of excellent tonal quality, making it easy to identify the people and locations they depict, which contributes to the process of active remembering (Wertsch 2002, p.17) the images provoke. The Collection is replete with many hundreds of images of actual forebears of the Lamalama, now also known to them as Old People, with whom they regularly interact, and to whom they must answer for their on-going behaviour.

The Lamalama find no discontinuity in this, despite the fact it requires them to view images of their Old People as they were in life. Aboriginal mortuary practices generally include a cultural prohibition on any mention of the deceased for a certain period after death. This does not seem to apply in the present context, perhaps because of the length of time between the death of the people in the images and their viewing by current Lamalama descendants (Hafner 2005), or perhaps because the people depicted in the images were never actually known to most of the Lamalama living today. Whatever the case, fear of any kind of retribution from the Old People for discussing or viewing images of the deceased does not seem to pertain to Lamalama participation in the research (Hafner 2005).

Rather, people now see it is part of their responsibility as owners of their country and custodians of its traditions to actively pursue this research into the Collection. Indeed, it is as a direct result of their requests that the research proceeds. In the following section, I describe this research, which demonstrated the way the information contained within the Thomson Collection acts to mediate a past that is useable in the production of current identity, through a process of collective remembering. The evidence in the Collection points to a particular past, known, believed in and to a lesser extent experienced by some of the Lamalama. Previously, this past was unavailable to most, except through the memory and associated story-telling or 'yarning' of older people.

Accessing and Using the Past

The Thomson photographs depict people engaged in a variety of activities, some as simple as watching over a child while it splashes in a river to more complex activities including mortuary rituals and other ceremonies.

They show men engaged in the use of large circular fishing nets (Figure 2), and women engaged in gathering plant foods. There is also a considerable number of portraits of individuals and small family groups, but it is not just the images of people or objects contained in the photographs that authenticate the past for the Lamalama.[8] Some of those who were small children at the time recognize actual individuals in the images. In most instances, though, it is the individual's presence in a group of people, or at a particular location within the Yintjingga region that produces an anecdote about the past that other Lamalama can respond to as authentic evidence of their history. In the current research, we therefore seek to access this process of collective recollection, which as Wertsch (2002) points out, might more accurately be referred to as one kind of 'collective remembering' or 'knowing' rather than collective memory.

Wertsch (2002, p.21) distinguishes between 'strong' and 'distributed' forms of collective memory. The former is associated with the idea that any social group can be regarded as having a memory in its own right, an idea sometimes attributed to Halbwachs (1992). Douglas (1980) saw this approach as a reification of collective memory into quasi-mystical status. Equally, Wertsch (2002, p.22) points out, Halbwachs argued that it is individuals as *members* of groups who remember, not the groups as entities in themselves. In contrast to such 'strong' forms of collective memory, Wertsch (2002, p. 25) describes distributed memory as the process of recollection that is spread across the members of the group, and distributed between them and the tools they employ to remember and carry out other actions in the world. There is no commitment to a collective mind in this form as there is in strong versions of collective memory (Wertsch 2002, p.22-23), and it is one form of this process that the Lamalama engage in when they view the Thomson materials. The kind of distributed collective memory that most fits with the Lamalama process is what Wertsch (2002, p.23) refers to as 'complementary distribution'. In this form, individuals within a group have different memories and perspectives that exist in a coordinated system of corresponding or complementary pieces of a larger system of memory.

In their interactions with the Thomson Collection, the Lamalama are engaged in a process of retrieving knowledge about the past that depends more on this kind of collective interaction than simply the direct memories

---

[8] The considerable size of the Collection and earlier constraints on its accessibility mean the Lamalama are now able to contribute their knowledge to its continuing curation.

of individuals. That is, they are engaged in piecing together clues from individual memory fragments, but also from a more complex body of information. This includes individual knowledge about genealogical history, clan affiliations, and the environment shared within the group, but it also includes a complementary process of remembering that relies on knowledge of events in the broader world to recall information about their own history. The Spanish Flu epidemic of 1918 and World War Two, for example, were historical events that Lamalama people have used to anchor personal memories. Thus Lamalama constructions of the past draw on their relationships at both a global and a local level. They draw on the memories of individuals, as well as what individuals know as a matter of transferred information – those narratives about past events passed down to them as a matter of oral tradition.

This process has already been apparent in the current research on the Thomson materials. In February 2005, three Lamalama people visited Melbourne to work on the Collection. During their visit they reviewed storage arrangements for archived materials, and provided information about selected images and artefacts to the Collection curators, all of which was videotaped. As part of their discussion, one of the young women present talked in detail about the activities of a group of women in one of the images. They were engaged in a process of food preparation which she had never seen performed herself, but she was able to provide considerable information about what was being depicted. This was because she had been told about this type of food and its preparation by older relatives as she grew up.

In April 2005, we returned copies of the images this group had worked on to the Lamalama in Coen for discussion with the wider group. We were testing the effectiveness of this as a methodological process, in the belief that distributed memory was more likely to emerge in the Lamalama home setting than at the Museum. We were able to engage a much larger number of people there in discussion of the images and what they revealed. On the first day, we met in the back yard of the senior man who had come to Melbourne, whose house occupied a central location in Coen and operates informally as a Lamalama 'drop-in' centre. People came and went as they chose, and a large number of children ran around, played, and looked at images with their older kin (Figure 3).

The images we worked with were all A3-sized black and white copies of Thomson's glass-plate negatives, which were quickly distributed among a group of mixed gender and age. During this session, one old man was disconcerted by the vulnerability of personal memory when he discovered he was looking at an image of his mother as a young woman. He did not recognize her until prompted by his wife, who reminded him of who he was looking at and where she was at the time the photograph was taken. It is possible this man would not have recognized her at all if his wife was not there to jog his memory. It is in larger groups of this kind, preferably on their own country, that this complementary process of distributed memory is most likely to take place. Because it was the monsoon season at the time of our visit and a large cyclone was threatening the region, we were unable to take the materials to Port Stewart and carry out the more detailed work of memory recovery we had planned. This would have involved using the images to replicate the activities they depict, and is among planned future stages of the research.

Nonetheless for us as researchers, both the process and the resulting information were important to record. We were especially interested in the way people across the group chose to view and discuss the images. Women from the senior Port Stewart land-owning clan discussed the images in both English and their own language, Ayapathu; the senior man and his wife described above sat together and discussed them only with each other, while younger people present engaged in more general discussion. Everyone present showed an intense interest in the materials, as Figure 3 demonstrates. The implications of these interactions remain to be tested further, but seem to indicate how intimate a process remembering can be, even when framed in a mutually shared and collective environment.

Importantly from our point of view, the presence of younger and older Lamalama people combines long-term memory and knowledge with an understanding of the technological possibilities offered by the Thomson materials. While viewing the images in Melbourne, one of the young women discussed ways they could be electronically manipulated to teach traditional practices and values to their children. These included using digitised versions of the images to develop computer games. As she commented, 'All our kids got mobiles [cell phones], all play them games. More better if we teach them one-time [about our culture], make our own games and teach them Lamalama way'. The digitisation of the Thomson images will further the goals of the Computer Cultures project as well, locating considerable Lamalama resources in one accessible form through the development of a computerised database. This of course happens in the context of an oral culture in which knowledge is passed down between the generations by the telling of stories that recount the exploits of the Story beings and the Old People, providing a moral universe that describes 'the Lamalama way'. An important part of the research will be to explore the degree to which interactions with the Thomson materials will work to reinvigorate those aspects of Lamalama practice that, in their own characterisations, have been 'lost' to them until now.

The research thus engages in a process by which the Lamalama past becomes an active element in the Lamalama present. In this process, digitisation, the most 'modern' form of transferring information, becomes the servant of tradition. The Lamalama clearly intend to use digital technology to extend their custom of oral transfer of information. It is unlikely to supplant the oral ways that tradition is now transmitted, such as the telling of stories about country by the old to the young, the censuring of inappropriate behaviour on country, and the sharing of information within the five major Lamalama families. But it undoubtedly will contribute to the ways in which the Lamalama carry out these activities. Just as generations of other Aboriginal people have also engaged in the practice of bricolage with regard to available tools and technologies, in the Lamalama view using electronic and digital technologies is simply harnessing the most effective means for continuing their traditions.

They are not concerned that employing digital technologies will destroy or take away from their traditions – it would be too late for such a concern anyway, as they have always enthusiastically embraced the advantages that supposedly non-Aboriginal technologies allow. Mobile phones are a great advantage to people living in remote locations who cannot depend on the regularity of a variety of services, from telecommunications to medical assistance. The Lamalama use such technologies for their own purposes, as they have other, previous waves of technology. Indeed, they see such technologies as mediating their attempts to maintain and extend their traditions in the face of severe social dislocations – alcohol abuse, unemployment and boredom pose much more serious threats to the continuation of their culture than current technologies.

Collective Remembering

Schacter (1996) points out that memory in the individual is not one single thing, nor simply a passive recording of past realities. Instead, as individuals we hold on to the 'meaning, sense and emotions these experiences provided us' (Schacter 1996, p.34) in the process that Tulving (1983) described as 'episodic memory'. This is the subjective form of memory that depends on specific context to allow us to recall events of the past that pertain specifically to our own experience and therefore define our individual lives (Schacter 1996, p.17). Such memories are also thought to occur in either 'field' or 'observer' mode (Schacter 1996, p.21). The

observer mode of remembering is associated with memory of the more
objective circumstances surrounding the recalled event, while the field
mode is more likely to include recollection of the individual as a
participant in the event recalled, including how he or she felt at the time
the remembered incident actually occurred.

From this perspective, a significant part of the recollection of events is
constructed at the time a memory is retrieved (Schacter 1996, pp.21-22),
so that our purposes in recalling an event inevitably help to frame the
memory that results. Thus the emotions that arise when we seek to retrieve
memories may at times be associated with the way in which we set out to
retrieve them. This helps to explain the surprised reaction of the old man
when viewing the long-forgotten image of his mother. His discussion with
his wife up to that point was not of a deeply personal nature, as he was
viewing images of people he knew about, but had not known personally.
When he realized that he was looking at his own mother, albeit at a time
before his birth, he moved from an observer to a field mode of
recollection, re-experiencing his emotions at her loss and, I believe, a
certain degree of embarrassment or anxiety that he had not immediately
recognized her himself. His reaction was one that mixed past and present
emotions around a memory triggered by visual recollection.

Schacter (1996) describes this kind of experience as the difference
between remembering and knowing the past. He points out that some of
the same regions of the brain are responsible for visual imagery and visual
perception, so that recall of visual information is taken to be crucial to
whether or not the individual is 'remembering' or 'knowing' the recalled
event (Schacter 1996, p.23). In the case of the old man, his wife was able
to 'remember' his mother and prompt his own memory. She was able to
use her recollected knowledge of the visual clues offered by the location in
which his mother was standing to retrieve her own autobiographical
memories of her.

However, Wertsch (1996, p.36) cautions against relying too heavily on
individual memory as a model for collective forms of rememberin. He
stresses that it could be quite misleading to assume that the specific
psychological and brain mechanisms on which models of individual
memory are based have their counterpart in collective memory processes.
Of concern here, however, is that there is very little autobiographical
experience of the events being recalled by the Lamalama to draw on.
Nonetheless memory is invoked when they view and discuss the people
and events depicted in the Thomson images. As Wertsch (2002, p.39)
points out, even direct autobiographical experience can be mediated by
linguistic and cultural experience, as when world events are experienced

through broadcast media rather than direct personal experience. Living through certain events may not be any more 'autobiographical' an experience, and therefore provide more direct memories, than experiencing them primarily through other media such as news broadcasts. This implies that in a culture of oral tradition, recollections based on oral transmission of information may be more directly experienced memories if they are associated with the autobiographical experience of being told about the people, places and events by individuals one directly knows or remembers.

Lamalama experience of collective memory in this context is one that relies less on content or 'episodic' than process or 'instrumental' memory of what is being remembered: that is, the memory about 'cultural tool[s] and the procedures for using [them]' (Wertsch 2002, pp.51-52). For example, specific language may be important in shaping the way a collective group remembers events, and in fact a range of cultural tools might be used to produce collective episodic representations of the past (Wertsch 2002, p.52). Similarly, the experience of remembering an event across cultures and epochs may vary: differing cultural tools and the ways in which they are used can be expected to yield varying results for separate collectives remembering the 'same' historical event. Yet Boyer (1999, cited in Wertsch 2002) argues that such transmission of information about the past may be the subject of universal cognitive constraints, so that certain concepts are more likely than others to be acquired by human groups, making some concepts 'much easier than others to acquire, store and communicate' (Boyer, cited in Wertsch 2002, p.55). Wertsch concludes that Boyer's position suggests the range of memories in both instrumental and episodic memory may therefore be limited by universal cognitive constraints.

In my view, however, this position does not preclude understanding Lamalama experience in relation to the Thomson images as one of distributed collective memory. The Lamalama use the Thomson images to re-animate their identity by using knowledge about the past that is contained within them, distributed in the memories of individuals and accessible as a set of cues that trigger fuller expressions of their own history. They draw on the cultural tools of oral tradition to share memories they have acquired through the important activity of 'yarning', to re-distribute knowledge and institute memory of the past in succeeding generations.

Concluding Remarks

For the Lamalama, the past does not exist as a set of memories retrieved through the narratives of oral tradition alone. The Thomson images form a partial visual record they can also draw upon to reinforce their current representations of the past. These representations include the cultural tools of electronic technologies now available to shape the understanding of their past being reconstructed with the aid of the Thomson materials. The question of whether the availability of more modern technologies to carry out the function of traditional artefacts makes such efforts redundant is not one the Lamalama pursue. Being able to produce a dug-out canoe as a result of exposure to the Thomson images is an end in itself; it is not important to them that an aluminium dinghy will do the same job.

Of greater importance is that young people participate in and learn the appropriate 'Law' about these activities and that they are videoed and a record of the event retained. In turn, the event and the film that results will be remembered, recalled, and enter into the body of cultural tools that forms the instrumental memory from which they construct a contemporary identity in the face of overwhelming stresses to give up traditional ways and enter the 'modern' world. In this they choose to confront the modern on their own terms, and appropriate it to the needs of tradition.

References

Allen, L. 2005, 'A photographer of brilliance' in B. Rigsby and N. Peterson (eds.), *Donald Thomson, the Man and Scholar*, Canberra: The Academy of the Social Sciences in Australia, pp.45-62.

Attwood, B. 1992, Introduction in B. Attwood and J. Arnold (eds.), *Power, Knowledge and Aborigines*, Special edition of the Journal of Australian Studies (35):i-xvi.

Australian Broadcasting Commission 2001, *Bush Mechanics*, Sydney.

Bartlett, F.C 1932, *Remembering: A study in experimental and social psychology*, Cambridge: Cambridge University Press.

Boyer, P. 1999, 'Human cognition and cultural evolution' in H. L. Moore (ed.) *Anthropological Theory Today*, Cambridge: Polity Press, pp. 206-233.

Bolton, L., 2003, 'The object in view: Aborigines, Melanesians, and museums', in L. Peers and A.K. Brown (eds.) *Museums and Source Communities: A Routledge reader*, London and New York: Routledge, pp.42-54.

Cape York Partnerships, 2005, 'Community Celebrates Indigenous Knowledge and Education', press release dated 22 June 2005, at http://www.capeyorkpartnerships.com/team/computerculture/news/cc_launch.htm, viewed 23 January 2006.

Commonwealth of Australia, 2002, *Connecting Regional Australia: The report of the regional telecommunications inquiry*, Canberra: Commonwealth Department of Communications, Information Technology and the Arts.

Douglas, M. 1980, 'Introduction: Maurice Halbwachs (1877-1945)', in M. Halbwachs, *The Collective Memory*, New York: Harper & Row (translated by Francis J. Didder, Jr and Vida Yazdi Ditter).

Hafner, D. 1999, *Feelings in the Heart: Aboriginal experience of land, emotion and kinship in Cape York Peninsula*, PhD thesis, University of Queensland.

—. 2005, Images of Port Stewart: Possible interpretations, in B. Rigsby and N. Peterson (eds.), *Donald Thomson, the Man and Scholar*, Canberra: The Academy of the Social Sciences in Australia, pp. 211-230.

Halbwachs, M. 1992, *On Collective Memory*, edited and translated by L. A Coser, Chicago: University of Chicago Press.

Morris, D. 1989, *Domesticating Resistance: The Dhan-Gadi Aborigines and the Australian State*, Oxford: Berg.

Rigsby, B., 2005, 'The languages of eastern Cape York Peninsula and linguistic anthropology', in B. Rigsby and N. Peterson (eds.), *Donald Thomson, the Man and Scholar*, Canberra: The Academy of the Social Sciences in Australia, pp.129-142.

Schacter, D. L. 1996, *Searching for Memory: the brain, the mind and the past*, New York: Basic Books.

Sharp, R.L., 1952, 'Steel Axes for Stone Age Australians', in E.H. Spicer (ed.) *Human Problems in Technological Change*, New York: John Wiley, pp.69-90.

Sutton, P. 1995, *Country: Aboriginal Boundaries and Land Ownership in Australia.* Canberra: Aboriginal History Monograph 3.

—. 1998, *Native Title and the Descent of Rights*, Perth: National Native Tile Tribunal.

Tulving, E. 19873, *Elements of Episodic Memory*, Oxford: Clarendon Press.

Wertsch, J.V., 2002, *Voices of Collective Remembering*, Cambridge: Cambridge University Press.

Zamora, L.P.1998, *TheUsable Past: The imagination of history in recent fiction of the Americas*, Cambridge: Cambridge University Press.

POST-SCRIPT

BIANCA MARIA PIRANI & IVAN VARGA

We have chosen this title because the breathtaking speed of advances in neurosciences, in various fields of technology, their applications and their interference in the human body prevents us from making but temporary conclusions. Moreover, the ever expanding sphere of the cyberspace influences the shape of social relations, processes of cognition as well as of social intelligence. We are just making the first steps in understanding the far-reaching consequences of these changes.

This book attempts by analyzing key areas of these processes and innovations to provide a better understanding of our changing social intelligence. The expertise of our contributors cover a wide range of areas thus opens up a dialogue across disciplinary frontiers needed for analysing the new conditions under which (at least but not exclusively) the developed societies live and towards which the global society moves. Also, this analysis is significant for having a better grip of how we produce knowledge and how the new technologies of knowledge production will change what we can know.

Science and technology account for many characteristics of contemporary – post-modern – societies: the uncertainty, unaccountability and speed that contribute, at the level of personal experience, to unexpected changes, to the reduction of individuals to classifications that establishes new varieties of social control; the oscillation between visions of doom and visions of progress that introduces ambiguity in perception of the natural and social world.

Both doing and being, whether in the high citadels of modernity or its distant outposts, are carried out in territories shaped by scientific and technological invention. It is difficult for social scientists to identify forms of human organization or behaviour anywhere in the world (except in remote and isolated communities) the structure and organization of which have not been affected, in various degrees, by science and technology.

The dynamics of politics and power as well as of culture are impossible to be taken apart from the broad currents of scientific and technological change. What we know about the world is intimately linked to our sense of what we can do about it, as well as to the perceived legitimacy of specific actors, instruments and courses of action. Whether power is conceived of in classical terms as the legitimate use of force or hegemonical means or else, in terms proposed by Michel Foucault, as decentred in society and implemented by many kinds of institutions and exercised through the disciplining the body, science and technology are indispensable for exercise of power. In short, science and technology operate as *political agents*.

It is impossible to write the political history of the twentieth century without taking into account its most salient techno-scientific achievements, such as the discovery of nuclear fission and its product, the nuclear bomb; of the genes and the DNA and genetic manipulation, including cloning; radio communication, television, powered flight, computers, microcircuitry and scientific medicine.

What happens in science and technology today is inter-woven with issues of meaning, values and power in ways that demand sustained critical inquiry. Social and political arrangements for exploiting, resisting or just accommodating technological change do not emerge, intact and fully formed, in response to innovation and discovery. Technology does not 'drive history'. Legal and political institutions lead, as much as they are led by, society's investments in science and technology. In engagements with the physical world, we are not mere spectators whose responses and destinies are ineluctably transformed by the growth of knowledge and the acquisition of novel technological capability. At the same time, when we tune into the rhythms of everyday life, even at times of exceptionally rapid techno-scientific change, we experience more often the continuity than the sense of disequilibrium. In short, the ways in which we take note of new phenomena in the world are tied to all points, like the muscles on a skeleton to the ways in which we have already chosen to live in. Yet, astonishingly, most theoretical explorations of how social worlds evolve only imperfectly reflect the complicated inter-play of the cognitive, the institutional, the material and the normative dimensions of society. Several decades of research in science and technology studies have done much to illuminate how orderings of nature and society reinforce each other, creating conditions of stability as well as change, and consolidating as well as diversifying the forms of social life. A compelling body of scholarship has demonstrated that science and technology can be fruitfully studied as social practices geared to the establishment of varied kinds of structure and authority (Biagioli, 1999; Jasanoff *et al.*, 1995; Pickering, 1995; Clarke and Fujimura, 1992).

So viewed, the workings of science and technology cease to be a thing apart from other forms of social activity but are integrated instead as indispensable elements in the process of societal evolution. Increasingly, the realities of human experience emerge as the joint achievements of scientific, technical and social enterprise. Not only the sameness but also the diversity of contemporary cultures are derived from specific, contingent accommodations that societies make with their scientific and technological capabilities.

The account of change that seems to have captured the public discourse about networks is one that suggests the change this technology instigates will be revolutionary in the truest sense: it is believed that networks will fundamentally alter relationships of power in society. The prophets of the 'net-work revolution' believe we are on the cusp of a new world in which the spark is replaced bit by bit. For instance, the mental and moral dimension of technological development can be studied from two aspects:

■ the effects of the spreading use of increasingly integrated and networked information technology on human interaction and the dynamics of economic processes;

■ the transformation of the world view by digital thinking, particularly evident in the rapid development of new, digitally oriented research areas.

We are necessary concerned with space and politics. The production of space is best known through the work of Henri Lefebvre (1991) who argued that space and spatial relations are the material and social outcome of capital. Pierre Bourdieu in many of his writings also emphasised that the social space individuals occupy (e.g. urban areas of dwelling) depends on their social status and the social capital they possess or lack. The totality of relations and practices between things and people, the discourses or representations of space, all act to produce space. They do not just act within a passive space but actively generate or produce real spaces and places. The subsequent explication of the production of cyberspace has been usefully initiated (Graham, 1998; Dodge and Kitchin, 2001) but the topic has remained relatively unexamined compared to the quantity of literature we find on the production of space more generally (e.g.Brenner and Elden, 2001).

We consider therefore how mapping of space and cyberspace constitutes a political engagement as well. This process includes the blurring of boundaries, in this case between physical and virtual space. Our inquiry specifically focuses on the new boundaries between bodies and technologies with the aim to find answers to the following questions:

1. What are the material relations of the production of cyberspace?

2. How is cyberspace extended by the relations of power/knowledge?
3. How is the social actor formed in relation to cyberspace?

In other words, we are interested in fully grasping theminterrelations between physical and virtual space. This approach we can call, after Foucailt (Foucault, 1997:114) one of 'problematization' where we see the development of 'domains, acts, practices, and thoughts that seem to pose problems for politics'. In other words, we are interested in fully grasping the interrelations between physical and virtual space. This approach we can call, after Foucault (Foucault: 1997a: 114), one of 'problematization' where we see the development of 'domains, acts, practices, and thoughts that seem to pose problems for politics'. By thus emphasizing such a problematic of cyber-space, we wish to signal the fact that cyberspace is an ongoing outcome of the material and virtual relations of production that the social actor finds him- or herself- a part of this production, and is in turn produced. But it is the producing which is important, not cyberspace as an end-product. We wish therefore to point out that it is a constant outcome of the material and the virtual relations of production[1] , that it is the social factor which makes the individual part of this production, and makes them produced. It is, however, in a socio-logical and socio-psychological aspect, the producing element that is important, and not the cyberspace as end-product. Nevertheless, as already Norbert Wiener warned, there is a difference between the producers and the users of cyberspace. In the late 'fifties he said: '[...] these computers will ruin the brains of all people. A few people will program them and the whole public will just mechanically follow.' (Wiener, in Conway and Siegelman, 2005:308).

More explicitly, by problematizing cyberspace, we mean the following:
1. It is to set something as a question, either by us, or more usually one which occurs in the discourses on the time and   place under study. Why and how is cyberspace at issue?
2. Second, to problematize something is to undertake a history of thought rather than a history of ideas and/or representations. Histories of cyberspace (e.g., Hafner and Lyon, 1996) tend to follow the progressivist chronological emergence of technical developments rather than examine how our relationship to technology has opened up or closed off meaningful thought and practice. An illustrative corrective is offered by Foucault, namely that 'it was matter of analyzing, not behaviors or ideas, nor

---

[1] This is not meant in the Marxian sense of the term. Rather, it signifies the relationships amongst individuals who create, respectively use, the technology that makes up the cyberspace.

societies and their 'ideologies', but the *problematizations* trough which being offers itself to be, necessarily, thought- and the *practices* on the basis of which these problematizations are formed (Foucault, 1985: 11).
3. Finally, to problematize is to examine the truth claims of the discourses: 'problematization' doesn't mean representation of a preexisting object, nor the creation by discourse of an object that doesn't exist. It is the totality of discursive or non-discursive practices that introduce something into the play of true and false and constitutes it an 'object for thought' (Foucault, 1988c: 257). With these three aspects of problematization in mind, we would like to attempt, if only in a preliminary way, a *spatial* problematization of cyberspace, or what we label 'being virtually there'. The advantage of this label is that it invokes being, presence-absence, digitality, spatiality and amalgams of these terms. Table 1 attempts to interpret these terms and to assign them a place.

Table 1. The problematics of the book: a genealogy of the social actor for cyberspace

Term	Topic	Enquiry
Being Authentication	Problematization	Problematization
Being Present-absent	Digital being, being Disciplinary cyberspaces	Confession and Parrhesia
Virtually divide	Almost, deferred, digital	Geoghaphy of the digital
Virtually there Why mapping is political	Digital spatiality	History of cyber mapping
Being virtually there	Ethos	An ethics

Our basic position is that the digital technology's important theoretical and cultural implications cannot be fully understood without examining both its materiality and immateriality. These are not contradictory qualities but rather mutually constituting elements. Cyberspace for us today has become the site of a number of problematizations and issues, ranging across debates over privacy and surveillance, access to technological resources and knowledge, the nature of relations between the physical and the virtual, identity and authenticity, and the nature of spatial relations in cyberspace.

There can be no doubt that the 'Digital Revolution' is going to change the way knowledge is gained and the way wealth is created. The integration of private communication and mass communication into a single 'machine' is being further advanced by the digitalisation of the electronic mass media. In this digital communications complex which may be described as a 'network' in the many meanings of the word the boundaries between private and public, local and global, and perhaps ultimately also present and absent, seem to be blurring as real-time presence.

The question that arises therefore is: are we content to hold these two contrasting 'attitudes' (in the physical/corporeal, as well as cognitive/sense, presence/absence, virtuality/reality)? It is certainly a tension that we find expressed in many fields where debates are held through such dichotomies as the real versus the constructed, the pure versus the hybrid, the essential versus the relational. We can point to a similar terrain on which are conducted disputes over nationalism and cosmopolitanism (Cheah and Robbins, 1998) and multiculturalism and hybrid identity politics (Werbner & Modood, 1997; Ratnum, 1999). We have used the term 'terrain' quite deliberately. Because it is a spatial metaphor, it is a space that can be traversed. It certainly is a messy space - with positions cross-cutting, hybridizing, purifying and essentializing all over the place. Moreover, it is a space that can be surveyed: the messiness are inflated by the constant claims to have mapped out the terrain to have sorted out and named the locations, ranked the positions, laid down the paths that connect them. Inevitably, such mapping claims become, in their turn, contested locations on the very terrain that was supposed to be mapped. Importantly, for our purposes, this messy spatiality is even messier when we begin to deal not only with positions made up by representations, discourses, and common places, but with co(a)gencies, that is, where positions reflect not only the histories of cultural resources (e.g. the differing epistemologies of realism and relativism) but also the differing lineages of technologies and natures. To this image of a messy spatiality we can add that of messy temporality. It is not that at certain times we have stable patterns of association, and at other times there is a disorder: these are not discrete times that can be arrayed, that is spatialized, along a time line. Rather, these sorts of interactional dynamics 'make time' in Serres's terms, different times emerge from these different sorts of configuring (Brown, 1999). Such 'emerging times' calls on the body to inform the concept of 'medium' and also to provide the potential for action *within* the space-time of information. The digital image is indeed configured as an autonomous *technical* image, one that carries out its work without any necessary correlation whatsoever to human perceptual ratios. If the digital image can be said to replace photographic, cinematic, and televisual images with the totally new technical image, it is because it fundamentally reconfigures the very concept of 'image', stripping it of a correlation-by-analogy with the human body and thus rendering it a purely *arbitrary construct*.

This changeable reconfiguration signifies 'the *new boundaries between bodies and technologies*'. These boundaries are fluid and continuously crossed in search of refinements of the body with the aim of overcoming the limits that nature or accidents of life have imposed to it. To escape to the limits that physicality has imposed to humankind and to project the body in a dimension that challenges the same death on the push of a turbocharged optimism: we are in presence of an endless move of the threshold 'over' the physical body that knows neither definitions nor limits. We face mobility at the scale of the body, for it is with the body that mobility starts. Between mobility as a core facet at the heart of modernity, and mobility as a threat to the kind of rationality that modernity signifies constitutes the basic conflict that informs modern thought: mobility as Becoming versus the 'stable and static' as Being. The 'nomadic body' is indeed the 'hero' of the so called 'striated space' 'sprawling temporary, shifting shanty-towns of nomads and cave dwellers scrape-metal and fabric, patchwork, of which the striatons of money, work, or housing are no longer even relevant' (Deleuze and Guattari, 1987: 481).

According to Turkle (Turkle 2004), technologies are 'evocative objects' by which the 'nomadic bodies' explore intimacy, emotional resonance, perception and dreams, the private world of fantasy and subconscious memories. The so called 'computer culture' is growing indeed, familiar with the experiences of passion and dependency. The 'affective computing' are becoming the new 'transitional objects' in a ghost of a baby blanket or rag doll. For example, one user says: 'I become my computer. It is not just that I remember people or know more about them. I fill invincible, sociable, better prepared. I am naked without it. With it, I am a better person'. (Turkle, Ibid.: 2). In recent years, the power of the 'transitional objects' is commonly seen in experiences with computer. Most recently, a new kind of 'computational objects' has appeared on the scene. 'Relational artifacts', such as robotic pets and digital creatures are explicitly designed to have emotive, affect-laden connections with people. Today's 'computational objects' present themselves as already animated and ready for relationship. In the role of 'self-object', the other is experienced as part of the self, thus in perfect tune with the fragile individual's inner state. Just as television today is a background actor in family relationships and a 'stabilizer' of mood and affect for individuals in their homes, in the near future a range of robotic companions and a web of pervasive 'computational objects' will mediate social lives. Even the computer interface encourages rethinking complex identity issues.

The 'computational objects' constitute the new 'low-level knowledge objects'. The 'knowledge objects' can change the meaning of 'objects' inasmuch as we shall have collected in one place all quantitative and qualitative information about an object. Using the principles of object-oriented programming, we need to develop objects of objects, a richer kind of metadata, which will contain key information concerning them. This quest to achieve objects of objects which contain information concerning all the physical and qualitative characteristics of the original is analogous to the quest for determining the structure of DNA and the mapping of nature in the human genome project. It is much more than just another cataloguing project. It is a quest, which will transform the very meaning of knowledge. On a seemingly quite different front, companies such as Autodesk have extended the notion of object-oriented programming to the building blocks of the man-made world through what they term industry foundation classes. A door is now treated as a dynamic object which contains all the information pertaining to doors in different contexts. Hence if one chooses a door for a fifty-storey skyscraper, the door object will automatically acquire certain characteristics which are very different from those of a door for a cottage or a factory warehouse. This is leading to a revolution in architectural practice because it means that architects designing buildings will automatically have at their disposal the 'appropriate' dimensions and characteristics of the door, window or other architectural building block which concerns them. This same technology can be used with very different consequences if one extends the concept of foundation classes to include cultural and historical dimensions. This transforms the meaning of knowledge. The modern 'temple object' centres knowledge on the fundamental significance of differences. Thus 'temples' gain universal value through the richness of their local variation. Knowledge lies not in recognizing how good a copy is but rather in how well it has created a variation on the theme. Hence knowledge of spatial location, the coordinates familiar to Geographical Information Systems (GIS), and Area Management/ Facilities Management (AM/FM) will also be an essential part of a 'knowledge object'.

The third dimension has many uses beyond producing such electronic copies of the physical world. Pioneers of virtual reality such as Tom Furness III, when they were designing virtual cockpits, realised that pilots were getting too much information as they flew at more than twice the speed of sound. The challenge was to decrease the amount of information, to abstract from the myriad details of everyday vision in order to recognise key elements of the air and landscape such as enemy planes and tanks. This means to use spatial arrangements of concepts in order to map problems identified and to visualise which subsets thereof were financed as research projects, which were solved in the sense of producing patents, inventions and products and which led to new predictions in the form of hypotheses and projections. This approach can in turn be generalised for purposes of better understanding the contributions of a group, a learned society, or even a whole culture.

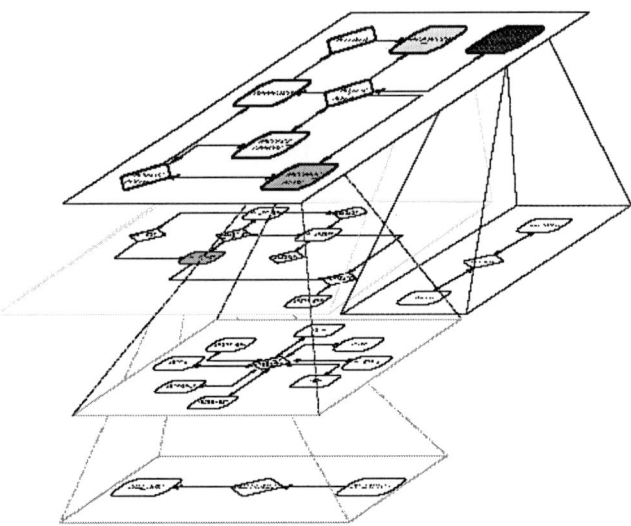

Table 2. Diagram relating to metadatabase research at Rensselaer Polytechnic in conjunction with Metaworld Enterprises entailing a Two Stage Entity Relationship (TSER) in the context of an Information Base Modelling System.

Synthetically, the main developments concerning the so called 'knowledge-based' societies are:

a) context (broadband technology allows an unpredictable development of application domains, some of them very risky concerning their ethical aspects);

b) paradigmatic shift (computing is seen as an interaction in open, dynamic, uncertain environments, rather than as a palette of services);

c) range (transferred from captological[2] applications to any agent based applications);

_____

[2] Captology is a word coined by Professor Fogg in 1996 from the initial letters of Computers as Persuasive Technology and adding the ending –ology.

d) approach (moving from persuasive agents to self-aware ones, the need of enhanced ethical control is growing);

e) software instruments (the design is based on agent oriented software engineering instead of generative programming);

f) key concepts (e.g. the syntagm 'digital ethics').

A basic assumption is that the 'knowledge-based society' whatever shape it would take entails new targets for e-Learning because:

a) humans must (inter)act in open, heterogeneous, dynamic and uncertain environments (OHDUE), quite different from the way they are familiar with.

b) The challenges to cope with are major and involve other requirements. Indeed, seen as resources, time is even more expensive while information tends to be almost free. Unluckily, that works for *static* knowledge (information as *noun*), not for *dynamic* knowledge.

'The network is the computer' we like to say. However, we can not abstract the networked computer as an instrument from the technological condition in which is situated. In social structural processes, we face real difficulties in understanding social change and diversity. Cultural specificity survives, for example, with astonishing resilience in the face of the levelling forces of modernity. Consider culture, in particular, or more accurately cultures.

There are multiple problems to interpret. They relate to the body-technologies relationship, to the place of time and space in human existence, to the body-mind as well as to the humans-machines relations and last but not least to ethical problems involving computer generated applications. Wiener, in his book *Human Use of Human Beings* but also in several other writings and speeches expressed concerns about the possibility of human beings controlling the computers (machines). Indeed, the ever increasing speed and complexity of tasks computers are capable to perform makes more and more problematic to control the correctness of data or solutions computers provide. With some exaggeration one could say that we have arrived to the point when a new book on the same topic could be titled '*Techno-use of Human Beings*' meaning that we are getting more and more dependent on the computers over whose programmes the average users have no control. Further, the problem of cyber-ethics is a growing concern. Several European and American universities held conferences dedicated to information ethics. The European Union established the European Group on Ethics in Science and Technologies (EGE) that investigates into the consequences of the proliferation of information and communication technologies (ICT) for social intelligence, the machine—human beings relationship, the ethical aspects of ITC implants for the human body, the cognitive and social impacts of artificial intelligence (AI), the impact of robotics and softotics (intelligent software agents) on labour and, in general, which is the most important for the topic of this book, the body-technology relationship.

As Rafael Capurro, a noted German ethicist remarked (Capurro, 2005)

> Information ethics arises [...] because this new medium created problems that could not be solved on the traditional basis of traditional rules and roles of hierarchical generation, distribution, storage and exchange of messages under the premises of mass media in democratic societies. What do truth telling or *parrhesia* mean in this new situation? [...T]he question underlying information ethics is [...] of a broader nature than the problems generated by the Internet. In this broader sense information ethics deals with questions of the digitaliziation, i.e. the reconstruction of all possible phenomena of the world as digital information and the problems caused by their exchange, combination and utilization.

There is, as Capurro noticed, a reductionist view of the human body as digital data. It is in contrast with the initial concept developed at the dawn of cybernetics as a new scientific discipline which was a result of close collaboration of mathematicians, engineers, physicists, neurobiologists, anthropologists and psychologists. They did not view the body, especially not the brain in terms of digital data but rather as an analogous process that has its own, specific algorithms. Wiener developed the idea that the feedback process is present in nature, in society and in the brain.[3] Thus began the transformation of the boundaries between bodies and technologies.

Stephen Toulmin, the British philosopher, in his article 'The Importance of Norbert Wiener', written shortly after Wiener's death (Toulmin, 1964), briefly traced the aftermath of the Cartesian mind—matter division. He reflects on La Mettrie's thought that 'a machine of sufficient complexity could reproduce any of those operations which, in Man, were declared to be characteristically "mental"'. (Ibid.) In order to avoid the contradiction in terms of the idea of a 'thinking machine', it was necessary to develop a systematic mathematical theory, capable to of representing [...] both the interactions of brain and limbs[4] involved in rational behaviour, and also the networks and linkages required in order to

---

[3] For a detailed description of the emergence of this collaboration and the resulting insights see Part 2, pp. 219—234 of Conway and Seligman, op. cit.
[4] In a way, the 'interactions of brain and limbs' was assumed by Marcel Mauss in his theory of 'techniques of the body'.

construct an artefact capable of simulating that behaviour; and […] it was necessary to build such an artefact'. (Ibid.)

Toulmin remarks that the technological achievement belongs to IBM while 'the more fundamental mathematical victory is owing above all to Norbert Wiener' (Ibid.) He explains his use of the word 'artefact' instead of 'machine' by saying that 'it is not just the technology of machines which has changed during the last three hundred years: our very concept of what constitutes a "machine" has been vastly expanded in the process, and as result the old philosophical questions take on quite a new look.' (Ibid.).

Wiener and scientists around him proved that

> A common pattern of ideas and inferences could be built up […] which was capable of elucidating both the design and operation of complex electronic networks and also the interconnections of the nervous system, the brain and the muscles. It was not a question of identifying brains and machines, or even explaining brain-processes on the analogy of mechanical processes previously understood.
>
> It was a question of working out a single system of concepts whose implications were capable of representing, equally well, the operations of the brain and the workings of a conceivable artefact. (Ibid.)

All these insights had a profound impact on the change in the boundaries between bodies and technologies. These became fluid by the developments in technologies that reveal, even if not yet fully, the working of the brain and aim at perfectioning the body. It has a positive effect, e.g. constructing artificial limbs connected to the individual's nervous system allowing him or her to fine tune those or else, to allow speech impeded persons to communicate, like the famous theoretical physicist Stephen Hawking. Thus the limits of the body imposed by nature or accidents are overtaken by technology. Similarly, mapping the DNA structure that made possible to discover genetic causes of certain diseases, and stem-cell research opens up possibilities for curing them. The negative aspect is the prospect of selecting favourable features for a future child, a 'made to order baby'. This creates serious ethical and social problems. Apart from the above sketched ethical problems of information technology (and we use the term in its broadest sense), advocates of genetic manipulation, like John Harris, a British bio-ethicist (Harris, 2007), argue that humans did use corrective devises, such as reading glasses, hearing aids, dentures, anti-inflammatory medications. Similarly, regenerative surgery helps to correct birth defects or injuries caused by accidents or war. Harris also argues that the individual has the right to use biotechnology to improve memory, intelligence and physical strength. His arguments can be thus summarized: we are in our right to replace with choice the limitations imposed by nature.

While just about all people wish a better, healthier and more satisfying life for themselves and for their children, the genetic enhancement - that is costly and would remain so in the foreseeable future - creates an even greater inequality not only amongst social classes but also amongst the population of rich(er) and poor(er) societies. Also, science is far from knowing possible adverse effects of genetic manipulation. (Even simpler procedures, as the use of the drug Thalidomide in the 1950s which was supposed to bring to the world 'stronger and healthier babies', produced catastrophic results.)

Another socio-cultural aspect of the problematic is that different societies have different views about what is health, whether longevity *per se* is desirable, the modes of curing diseases, what constitutes natural, and so forth. Western or westernizing societies increasingly blur, in theory as well in practice, the dividing lines between nature and culture - and technology is part of the culture - thus between body and technology.

Further, social consequences of the ever broadening utilization of communication and information technology include that information became commodity. True, some, and growing amount of programmes (like Linus, Wikipedia or YouTube, etc.) are free, even though not advertisement free, and distribution of information became less hierarchical. Nevertheless, the warnings of Norbert Wiener did not lose their importance. Privacy issues emerge with greater frequency and techniques of surveillance are rapidly developing. Governments as well as commercial enterprises collect more and more data about the individual who, in spite of certain regulations, has basically no control over their use. The proliferation of use of robots in manufacturing requires new skills and influences the labour market.

The advent of quantic and three-dimensional computers, the great advances of biology, neurosciences and the instantaneous communication opened up new fields in human knowledge. It is an exciting time in the study of the human brain and mind. Nary a week goes by that doesn't see a new discovery in the fields of neuroscience and psychiatry, giving new insights into human mental phenomena and mechanisms. Much of this advance in knowledge is the result of technological advances in brain imaging. It seems that almost every time one hears about some neurological experiment or advance in human neuroscience, the term 'brain scan' appears. Five of the most important technologies that have allowed scientists to peer into the workings and structure of the living human brain are: EEG (electroencephalograph); CAT (Computerized

Axial Tomography); PET (Positron Emission Tomography); MRI and fMRI (Magnetic Resonance Imaging) and (Functional Magnetic Resonance); MEG (Magnetoencephalography). As Gazzaniga rightly cautions, (Gazzaniga, 2005:198) modern neuroimaging techniques read brains, not minds. Neuroscience has discovered that the structure of our brain, with its neurochemical processes makes it 'sociable', that is, in all human relationships, be they rational or emotional, there is a brain-to-brain interaction. Neurosociology, for instance, discovered certain chemical processes in the infant's brain that is influenced by the infant-mother interactions. That *neural bridge* creates a mutual influence of the brains, as well as the bodies, of everyone with whom we interact. The conventional wisdom had previously viewed this bridge as a *disembodied* and often *unruly phenomenon.*

The nature of representation has been a core concern of science and technology since the earliest attempts to understand knowledge in social terms. When we review the ontological question 'what is real' we are immediately in the dilemma of everyone having his own perceptions and reasoning and all of us living in 'parallel worlds' without any connection, without any exchange and reinforcement/acknowledgement between them. This has lead into the dead-end of philosophical post-structuralism, which cherishes the individual's physical and mental 'solitude' and loneliness which finally ended up into our present problems with the definition of cyberculture and virtual reality: open-ended universality which can not be managed or overviewed, which is a meaningless, contextless chaos, an abstract place in which individuals and societies can only get 'flooded' with the noise of nonsense, and so submerge and succumb to it. How to overcome this loss of ground and orientation? One way is to stay just with the physical reality (*res extensa*), with the world of objects and physical extension we can jointly address, measure, and touch. The other, according to Karl Popper, is to form an 'objective reality' by forming a framework and adding pieces, like in a puzzle. We can say in the end 'objectively' that the puzzle is complete, if all parts are integrated and have their place. This approach is nicely complemented by Gregory Bateson who considers differences having an 'objective' character the moment we can jointly discriminate or inspect them.

In his new book *Social Intelligence: The New Science of Human Relations* (2006), Daniel Goleman demystified this loss of ground by suggesting how we can consciously use the biological thermostat as a force to enhance the quality of our life. What occurs *between* can be thought of as the range of relationships that exist within a social continuum. The relationships in the center of this continuum imply a close empathetic human relationship that is temporarily or permanently tuned to the experiences, needs, and feelings of another person. Goleman suggests that we are constantly involved in both close and detached relationships, and that our relationship with a person can appropriately shift back and forth between close and detached, depending on the circumstances. He says that our brain is designed to make connections with other humans, and that our relationships have a real biological impact.

The social brain refers to the very extensive circuitry active in some way during a social interaction. The social brain doesn't just take in what the other person is doing. It tells us what to do next to keep things operating on track. If we're upset or agitated and we're with the person, we might say something artfully because our social brain is telling us how to do it. But without it on-line, it lets us do whatever we want and sometimes with unfortunate consequences. Correlations between human cognitive processes and functions and decisions related to perception, attention and mnemonics have a decisive role in the establishment of social behaviour. By connecting neural processes with mental processes, Donald Hebb in his well-known study *The Organisation of Behaviour*, demonstrates how 'experience shapes the brain' (Hebb, 1949). Our relations mould not just our experiences, but our biology. A multilevel approach to the study of mind follows from the fact that the brain does not exist in isolation; it is part of the physiological-biological structure of the body. This is a prime example for the need to distinguish brain and mind. Descartes introduced a reduction of truth to what can be clearly and distinctly apprehended by mind, employing a method that is constructed from its own resources exclusively. Mathematics meets these requirements in that its objects are transparent and their generation the result of rules that are clearly and distinctly intuited. The result was not only a grounding of the unity of mathematics in a thinking actor but, with this an elevation of the mind, in contrast to the body, to what is most essential to us as persons forming together the current 'social intelligence': the crucial role played by the spatial and temporal flows which, through biological rhythms, connect the brain with the environmental limit.

According to Damasio (Damasio, 2004:314), the somato-sensory mass of the brain builds up the connections which the body's confines compound with the sensory-motor integration many aspects of behavior, motor actions and sensory processsing are inseparably linked to environment by means of neural activity maps coordinated in time. In most sensory systems, neurons form a map of sensory space through the organized projection pattern of their axon terminals. One consequence of this anatomical organization is that the ensemble response of a population of neurons to a particular sensory stimulus will be represented as a unique spatio-temporal pattern of activity within that region of the brain. In order to understand how these spatio-temporal patterns of activity emerge from the ensemble activity, and how the information contained in the patterns is accessed and encoded by higher levels of the nervous system, the functional architecture of the constituent neurons must be determined. This determination is topographically acted in context by keeping stability in body space. Self—other differentiation is fundamental for any sensor. The contribution the body makes to the brain is not limited to supporting vital operations, but includes regulating the space and time which organizes the contents of a normal mind. What we learn from this is the concept of joint comparison, which might not lead always to global 'objective' results but at least helps to come to shared assessments even among our diversity.

The primary purpose of perception is to enable interaction with the environment. Interactions may include walking from one place to another, grasping an object, talking to a person, or navigating a car. In return such actions directly affect our perceptions of the world. These interdependencies between action and perception are illustrated by the 'Action/Perception-Cycle' which explains mapping as a mechanism providing a spatialized acting of the world (see table 3 below).

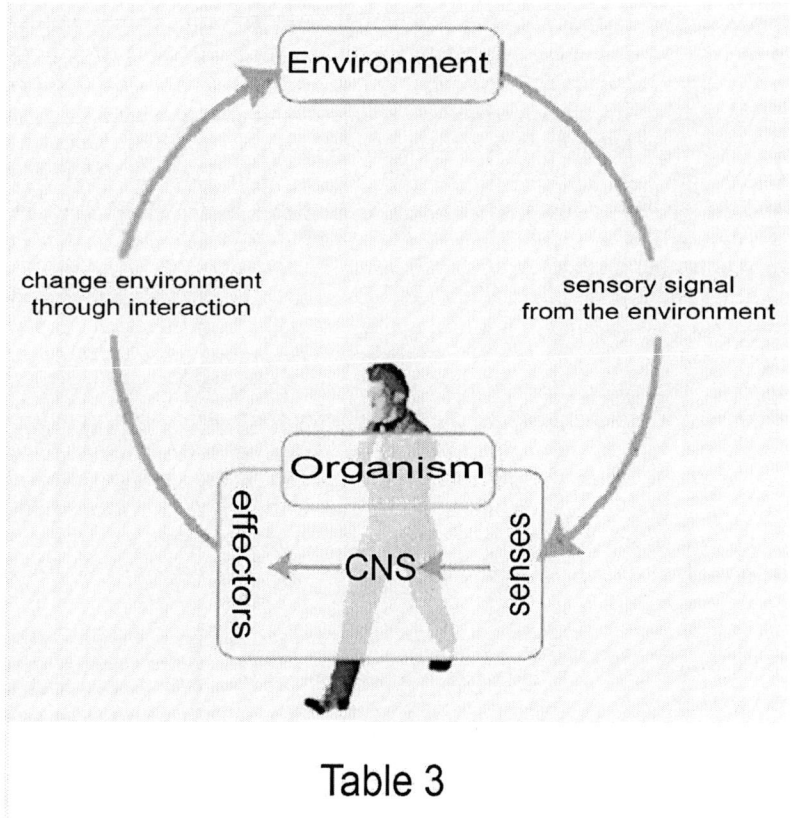

Table 3

In sensory-motor integration many aspects of behaviour, motor actions and sensory processing are inseparably linked in a closed action/perception loop. This is the basis of the idea to define common frames of references in an artificial agreed upon space. Consider it an imaginary space, a space-scape, a virtual reality, a place people can jointly immerse, encircle, focus on, pan and zoom, and outline themes and issues (topic-theme-scape) like you would consider a topos in the physical world (landscape). The Max Planck Institute for Biological Cybernetics of Tübimgen (and specifically the Department 'Cognitive and Computational Psycho-physics') has investigated the visual and haptic recognition of objects, orientation and navigation in threedimensional space, and the information processing underlying both. One question lies at the core of both of these topics: How are shape and space represented in the brain so that we can name and grasp objects or move in a given space (such as a unfamiliar town)? The use of computer graphics and 'virtual reality' makes it possible to carry out interactive experiments with human volunteers under strictly controlled conditions. The results have already been applied in machine vision systems in the automatic synthesis of faces and facial expressions and in automatic course-control of autonomous robots. To study sensory-motor processes the Researchers frequently employ modern high-fidelity Virtual Reality (VR) technology. That is, they replace the real environment in the action/perception-cycle by a virtual world. For example, to simulate objects which also can be touched, they have recently constructed a novel VR setup. At its core, this setup has two force feedback devices (PHANToM™) which enable them to generate arbitrary dynamic objects for the subjects to manipulate. These objects can have different haptic textures, masses, or degrees of stiffness. To allow for maximal interactivity, however, they need a different simulator for each sensory modality. With the motion platform, for example, they can stimulate the vestibular system. With the VR bicycle on the other hand they stimulate the proprioceptive[5] system and so mediate a sense of effort while they are actively exploring some virtual environment. This entire endeavour follows the sensory-motor philosophy which says that perception cannot be adequately studied in isolation from action processes.

Successful interaction with our environment relies on concurrent input from different senses. How does the brain learn to make specific predictions about the environment from its sensory inputs? What are the principles that govern how the visual pathways make inferences from the visual image? How do we use image information to compute these perceptual inferences? A principal difficulty in the understanding of biological vision is the complexity of the inference problems we encounter both at the level of behaviour as well as at the level of neuronal responses. This complexity mostly results from the large number of degrees of freedoms in the sensory input and in the neuronal responses. With the ability to send attribute data from personal information devices worn on the body to computers embedded in the environment, one-to-one services could be implemented that are tailored to the individual needs of the user. A major bottleneck in obtaining high performance with human—machine systems is the design of the human interface. Even the highest performance hardware and software can be seriously limited if the human

---

[5] Proprioceptive is a sensory receptor, found chiefly in muscles, tendons, joints and the inner ear *(vestibulum)*. It detects the motion or position of the body or a limb by responding to stimuli arising within the organism.

operator must work slower than the machine does. Thus designing human interfaces for systems such as airliner cockpits and computer user interfaces that maximize the total system performance is critical to the future success of our rapidly evolving technology. A brain-computer interface (BCI), sometimes called a direct neural interface or a brain-machine interface, is a direct communication pathway between a human or animal brain (or brain cell culture) and an external device. In this definition, the word brain means the brain or nervous system of an organic life form rather than the mind. Computer means any processing or computational device, from simple circuits to silicon chips (including hypothetical future technologies such as quantum computing). Research on BCIs began in the 1970s, but it wasn't until the mid-1990s that the first working experimental implants in humans appeared. Following years of animal experimentation, early working implants in humans now exist, designed to restore damaged hearing, sight and movement. The common thread throughout the research is the remarkable cortical plasticity of the brain, which often adapts to BCIs, treating prostheses controlled by implants as natural limbs. With recent advances in technology and knowledge, pioneering researchers could now conceivably attempt to produce BCIs that augment human functions rather than simply restore them.

How can we define mind? For 'mind' we usually mean the aspects of intellect and consciousness manifested as combinations of thought, perception, memory, emotion, will and imagination. According to this definition, mind is the stream of consciousness and includes all of the brain's conscious processes. Today, we need to discover that although the world may be shrinking in terms of physical access, its horizons continue to expand across the distributed assemblages in context creating mobile boundaries between sensory experience and technologies.

Before we turn to theoretical issues, we must not lose sight of one set of facts: the brain is embodied and the body is embedded. First, consider embodiment. All the brain activities depend on signals to the brain from the body and from the brain to the body. The brains maps and connections are altered not only by what you sense but by you move. As Edelman and Tononi have shown (Edelman, Tononi 2000:75-92), conscious experience is associated with the activation of neuronal groups, widely distributed in the cortical-thalamus system. Synchronism between the elementary self-consciousness which regulates somatic processes and the interaction taking place is responsible for the unrepeatable activation that produces the state of active consciousness. Second, consider embeddedness. Your body is embedded and situated in a particular environment, influencing it and being influenced by it. This set of interactions defines your 'econiche' as the changeable combination of bodily confines and cultural practices. The development of the whole person-organism is carried out by these boundaries, at the same time body and mind, situated within a continuous relational field. The flows at issue make up the topographical map-making of bodily experience.

Mapping is a fundamental capacity for the mental world. Without it every object and problem would be completely new to us and we would not be able to make use of any of our previous experiences to deal with them. As Herbert Simon's work (Simon, 1996) has shown, the mechanism of recognition is the main device of *problem-solving* capacity in present day experience. According to a computational view of the mind, mental maps, mind maps, cognitive models, or mental models are a type of mental processing (cognition) composed of a series of psychological transformations by which an individual can acquire, code, store, recall, and decode information about the relative locations and attributes of phenomena in their everyday or metaphorical spatial environment. Tolman (1948) is generally credited with the introduction of the term 'cognitive map'. Here, 'cognition' can be used to refer to the mental models, or belief systems, that people use to perceive, contextualize, simplify, and make sense of otherwise complex problems. Cognitive maps have been studied in various fields, such as psychology, education, archaeology, planning, geography and management. As a consequence, these mental models are often referred to, variously, as cognitive maps, mental maps, scripts, schemata, and frames of reference. Put more simply, cognitive maps are a method we use to structure and store spatial knowledge, allowing the 'mind's eye' to visualize images in order to reduce cognitive load, and enhance recall and learning of information. This type of spatial thinking can also be used as a metaphor for non-spatial tasks, where people performing non-spatial tasks involving memory and imaging use spatial knowledge to aid in processing the task. Maps are shortly 'workbenches of the mind' in the sense that each different map is, as Baars (Baars, 1997: 95) observes, 'the selective act that results in a conscious event'. The map contextualizes the ability of social actor to search locations, to find shortcuts and novel paths, to estimate distances between remembered places, and to draw sketch maps of the explored environment. Historians and anthropologists have now examined a wide set of cultural contexts in which mapping has been practiced and have documented a historical range extending into prehistory and across the vast majority of human cultures. Mapping practices are indeed contingent intersections of technologies and cultures and processes of social life. Maps preceded both language and counting. Each mode of mapping is intimately tied to social, cultural and technological relations, which are contingent on particular times and places. Distributed mapping is an emerging area that represents one of the most interesting outcomes of the convergence of spatial technologies such as GIS, remote sensing, and digital cartography with World Wide Web. This convergence combines the methods and techniques of interactive mapping and spatial analysis with the distribution of functionality and resources in new ways. Mapping is source of the production of geographic knowledge. According to Foucault there is no knowledge outside of power relations, and no power relations that do not invoke and create knowledge. Interests the mapping serves, as well as the 'knowledge objects', a history of ethics, are understood as the 'practices of the self' that Foucault (Foucault, 1999:181) calls 'politics of ourselves'.

We are indeed a tool-making species. And tools have always shaped the mind. We are not the only tool-making species, and we are not the only species whose 'cognitive life' is shaped by tools. Each new tool, whether it's fire or television, has shaped the mind of the humans who used it. Tools have 'created' the mind as it is now. That includes our 'artistic' expressions, that are part of a mental process of using the tools that previous generations have created. Where does the mind come from? According to the Darwinian paradigm, it must have evolved from something primordial and it must have evolved because it had a useful function for survival. The most primordial mind one can think of is a mind that has very basic emotions, possibly only pain and pleasure. There is mounting evidence that emotions have evolved, as much as any other organ. Emotions had an evolutionary value, as they helped bodies survive, and therefore were valuable, and therefore evolved. It is unlikely that humans are the only species with emotions, but it is likely that humans are the species in which emotions evolved in the most spectacular way. The reason for this spectacular evolution may very well be that we developed more sophisticated tools than any other species. Tools relieved us from many daily chores. Our emotions had been invented to help cope with those chores, but suddenly they were not necessary anymore. Our mind was nonetheless still producing emotions, just like the immune system is producing antibodies all the time. Those emotions flowing through our mind eventually got organized, and yielded thought. Thought eventually yielded a continuous flow of emotions and a concept of the self: consciousness was born. Consciousness was born because our mind had nothing to do most of the day, and that happened because we invented tools. As Leroi-Gourhan points out (Leroi-Gourhan, 1983: 35), 'the very problem is the *link* that connects the technologies to the brain of mankind'.

We can identify this link with the dynamic state of activity (number and frequency of impulses, free concentration of neurotransmitters, and so on) that in turn is correlated with a topologically defined and distributed neural connection by the acting of a given technical procedure. All the technical resources of human inventive capacity, from the chipped flints of the Neolithic Age to the Renaissance, respond indeed to the primary biological need to dominate the environment, something which is inseparable from human nature. In the classical notion, technology is basically understood as referring to the physical objects or artifacts produced by technical procedures. In that way, classical sociology of technology focuses only on the social use of those objects and obliterates the complex social construction involved in the material artifact, because classical sociologists of technology supposed that technology is the outcome of a process triggered when a group of scientists making basic research reveals the pattern that govern the natural world, and once appropriated by industry, this scientific knowledge translates into technology. So, it is assumed that there exist an univocal correspondence between scientific knowledge, technological production and innovation. According to this hypothesis, technology has been reduced to computers, consumer goods, and military weapons. We speak of 'technological progress' in terms of RAM and CD-ROMs and the flatness of our television screens.

We have to recall that the original word for 'technology' is the Greek *techné* which meant art, skill, way of making and doing. Technology and creativity are the same thing. And we are both the creators and the products of our technology. In today's context by technology we mean primarily mapping as ways of interacting that depend on instruments and digitality as a means. What we are interested in with technology, is how it can bring about insight into meaningful human life.

The authors of the book took up the challenge to be informative, critical and appreciative at the same time. They have contributed to initiate further analysis of the enormous complexity of social, scientific and cognitive problems arising from the current shifting boundaries between bodies and technologies, between humans and machines and between nature and culture.

References

Baars, B. J. (1997), *In the Theater of Consciousness*, New York: Oxford University Press

Biagioli, M.(1999), *The Science Studies Reader.* New York: Routdledge

Brenner, N. and Elden, S. (2001), 'Henri Lefebvre in Contexts: an Introduction', *Antipode* 33: 763: 768

Brown, N. (1999), Xenotransplantation: Normalizing Disgust. *Science as Culture*, 8, 327-355

Capurro, R. (2005), *Toward an Ontological Formation of Information Ethics.* http://www.capurro.de/oxford.html

Cheah, P. and Robbins B. (eds) (1998), *Cosmopolitics: Thinking and Feeling beyond the Nation.* Minneapolis: University of Minnesota Press

Clarke, A. and Fujimura J. (eds) (1992), *The Right Tools for the Job: At Work in the Twentyeth century Sciences.* Princeton: Princeton University Press

Conway, F. and J. Segelman (2005), *Dark Hero of Information Age: In Search of Norbert Wiener, the Father of Cybernetics.* New York: Basic Books

Damasio, A. (2004), *Looking for Spinoza. Joy, Sorrow and the Feeling Brain.* New York: Vintage

Deleuze, G. and Guattari, F. (1987), *A Thousands Plateaus: Capitalism and Schizofrenia.* Minneapolis: University of Minnesota Press

Dodge, M. and Kitchin R. (2001), *Mapping Cyberspace.* London and New York: Routledge

Foucault, M. (1985), *The Use of Pleasure.* Volume II of *The History of Sexuality.* New York: Vintage Books

—. (1988), 'The Concern for Truth'. In L.D. Kritzman (ed.), *Politics, Philosophy, Culture: Interviews and Other Writings, 1977-1984.* New York: Routledge, pp. 255-267

—. (1997), 'Polemics, Politics and Problematizations: an Interview with Michel Foucault'. In P. Rabinow (ed.), *Ethics, Subjectivity and Truth. Essential Works of Foucault 1954-1984.* New York: New Press, pp. 111-119

—. (1999), *About the Beginning of the Hermeneutics of the Self.* In Carrette (ed.), *Religion and Culture.* New York: Routledge

Edelman, G.M. and Tononi, G. (2000), *A Universe of Consciousness. How Matter Becomes Imagination.* New Yoork: Basic Books

Gazzaniga, M. (2005), *The Ethical Brain.* New York/Washington, DC: Dana Press

Goleman, D. (2006), *Social Intelligence: The New Science of Human Relationships.* New York: Bantam Books

Graham, S. (1999), 'Geographies of Surveillant Simulation' in M.Crang, P.Crang and J. May (eds), *Virtual Geographies. Bodies, Spaces and Relations.* London and New York: Routledge

Hafner, K. and Lyon, M. (1996), *Where Wizards Stay up Late at Night: the Origins of the Internet.* New York: Simon & Schuster

Harris, J. (2007) *Enhancing Evolution: The Ethical Case of Making Better People.* Princeton, NJ: Princeton University Press

Hebb, D.O. (1949), *The Organization of Behaviour*, New York: John Wiley

Jasanoff, S. (1995), *Science at the Bar: Law, Science and Technology in America.* Cambridge MA: Harvard University Press

Lefebvre , H. (1991), *The Production of Space.* Oxford: Blackwell

Leroi-Gourhan, A. (1983), *Le Fil du Temps. Etnologie et Préhistoire.1935-1970*, Paris: Fayard

Pickering, A. (1995), *The Mangle of Practice: Time, Agency and Science.* Chicago: University of Chicago Press

Ratnum N. (1999), 'Chris Ofili and the Limits of Hybridity'. In *New Left Review* 235: 153-159

Serres M. (1997), *The Troubadour of Knowledge.* Ann Arbor Mich.: Michigan University Press

Simon H.A., (1996), *The Sciences of the Artificial.* Cambridge (MA), MIT Press

Tolman E. C. (1948), 'Cognitive Maps in Rats and Men'. First published in *The Psychological Review, 55(4),* 189-208

Toulmin, S. (1964), The Importance of Norbert Wiener. *The New York Review of Books*, Vol. 3, No. 3, September 24, 1964

Turkle, S. (2004), 'Whither Psychoanalysis in Computer Culture?', in *Psychoanalitic Psychology: Journal of the Division of Psychoanalysis, American Psychological Association*, Winter 2004

Webster, A. (2002), 'Innovative Health Technologies and the Social: Redefining Health, Medicine and the Body', in *Current Sociology*, Vol. 50, No. 3, May 2002, 443-457

Werbner P. and Modood T. (eds), (1997), *Debating Cultural Hybridity*, London: Zed Books.

CONTRIBUTORS

Pierre Bouvier is Professor of Sociology in the Department of Sociology at the University of Paris X-Nanterre, researcher at the Laios/Ehess. His work focuses on social bonds and colonial/postcolonial issues in an interdisciplinary approach. He has founded the journal 'Socio-anthropologie" and has published numerous books, most recently 'La socioanthropologie' (Paris: Armand Colin 2000, re-ed. 2002) and 'Le Lien Social,Paris:Gallimard,2005, re-ed.2007).

Roberto Cipriani is Professor of Sociology and chairman of the Department of Sciences of Education at the University of Rome 3, Italy. He has served as a visiting professor at the University of São Paulo in Brazil, at the University of Buenos Aires, and at Laval University in Quebec and has conducted research in Greece and Mexico. He is past president of the Italian Sociological Association. His publications include: *Sociology of Religion. An Historical Introduction*; *The Sociology of Legitimation*; *"Religions sans frontiers?" Present and future trends of migration, culture, and communication*; more than 600 articles, and 40 books.

Carlo Donolo is Professor of Advanced Sociology – Analysis of Complex Systems and Institutions, Department of Social Accounting and Analysis of Social Processes, Faculty of Statistics, University 'La Sapienza', Rome Chief Research Interests: Institutional Analysis, Governance, Commons and Sustainable Processes, Policy Making and Policy Analysis.

Diane Hafner, PhD is a Lecturer in the Behavioural Studies Program at the University of Queensland, Brisbane, Australia. Her research interests include Australian Indigenous studies, identity formation, the social contexts of memory, and museum and curatorial practices. Her current research on memory and museum practice continues her long-standing work with Cape York Peninsula Aboriginal people.

Céline Lafontaine is an Assistant professor in the Department of Sociology at the University of Montréal. She holds a PhD in Sociology from the Université Paris 1 (Sorbonne-Panthéon) and the University of Montréal, and is interested in issues relating to the informational paradigm, technoscience and postmodern culture. Notably, she has published 'L'Empire Cybernétique. Des Machines à penser à la pensée Machines' (Éditions du Seuil, Paris, 2004).

Lauren Langman is a professor of sociology at Loyola University of Chicago. He got his phd from the university of Chicago, and psychoanalytic training at the Chicago Institute for Psychoanalysis. He work has been in the tradition of the Frankfurt School of Critical Theory. He is president of RC 36, Alienation Theory and Research of ISA, on the board of RC 48, social movements. Most recent book in on the Carnivalization of America

Giuli Liebman Parrinello, is Professor of Sociology at the University of Roma Tre, Teacher of Master in Linguaggi del Turismo e Comunicazione Interculturale. In the field of tourism studies her research interests and publications focus on tourist motivation, the evolution of post-industrial tourism, the impact of technologies (including cybertechnologies), the importance of the mind-body debate for tourism and on more theoretical and paradigmatic issues.

Tatiana Mazali has a Phd in Communication Sciences at the University of Turin. She is a researcher in the context of Sociology of Communication, Knowledge and Culture. She is lecturer at the Polytechnic of Turin in Cinema and Communication Engineering. Among his most recent publications: 'Ex-Peau-Sition, il corpo riscritto nelle performance artistiche', published by Guerini & Associati, Milano, 2007.

Umberto Melotti is professor of political sociology at the Faculty of Sociology of the University of Rome "La Sapienza". He has published widely on migration to Italy and Europe. His works in English include Marx and the Third World (Macmillan, London, and Humanities Press, Atlantic Highlands, N.J., 1980; also available in Italian, Spanish and Chinese) and his contributions to Sociobiology of Ethnocentrism (edited by V. Reynolds et al., Croom Helm, London, and Georgia University Press, Atlanta, 1987) and The Politics of Multiculturalism in the New Europe (edited by T. Modood and P. Werbner, Zed Books, London - New

York, 1997). He was also one of the main contributors of the Dictionary of Race, Ethnicity and Culture (Sage, London). In Italian he has recently published Migrazioni internazionali, globalizzazione e culture politiche (Bruno Mondadori, Milano, 2004).

Roberto Motta is former researcher at Fundação Joaquim Nabuco, Recife, and Full Professor at the Social Sciences Departmnent of Universidade Federal de Pernambuco, also in Recife. Nowadays he teaches at Universidade Estadual da Paraíba (Brazil) and is a researcher of Conselho Nacional de Pesquisas (CNPq), Brasília. He is the author of many published papers in several languages, bearing on the Afro-Brazilian religions, religion and development, race relations and Brazilian social thought.

Michèle Robitaille is a doctoral student in the Department of Sociology at the Université de Montréal. She is preparing her doctoral thesis on the representation of the body in the transhumanist movement.

Amelia Richards has obtained a masters degree in Research Psychology in 1998 from the University of Pretoria, where she continued her research and studies and obtained a PhD degree in 2006 titled: Generation X people's development of cyberspace culture: a psychological perspective. For the past 10 years she has completed research projects within the broad field of social sciences, consumer and market research with a specific focus on Internet research and Cyber Psychology from 2000 onwards.

Jeff Vass, Division of Sociology and Social Policy, School of Social Sciences, University of Southampton, Southampton. Author of Vass, J (1999) 'Social theories of the human agent and monastic dialogue' in Sociology and Theology in Dialogue (Francis, L,J) London, Cassell
Vass, J (1995) 'Economic socialisation: a tourist in my own transactions' Research in Economic and Social Understanding Vol 5 No 1